MIAMI IN THE ANTHROPOCENE

MIAMI IN THE ANTHROPOCENE

Rising Seas and Urban Resilience

STEPHANIE WAKEFIELD

University of Minnesota Press
Minneapolis
London

Portions of the introduction and chapters 4 and 5 appeared in a different form as "Critical Urban Theory in the Anthropocene," *Urban Studies* 59, no. 5 (2022): 917–36; copyright 2022 Urban Studies Journal Limited / Sage Publications. Portions of chapter 1 appeared in a different form in "Miami Beach Forever? Urbanism in the Back Loop," *Geoforum* 107: 34–44; copyright Elsevier 2019. Portions of chapter 4 appeared in a different form in "The Possibility of Islands in the Anthropocene," *Dialogues in Human Geography* 11, no. 3 (2021), and in "Urbicide in the Anthropocene," from *Imagining Apocalyptic Politics for the Anthropocene*, edited by Doug Specht and Earl Harper; copyright 2022, reproduced by permission of Taylor and Francis Group, LLC, a division of Informa plc. Portions of chapter 5 appeared in a different form in "Anthropocene Destitution," *South Atlantic Quarterly* 122, no. 1: 121–36; copyright 2023, Duke University Press, all rights reserved, republished by permission of the publisher, www.dukeupress.edu; and in "Anthropocene Hubris," *Accumulation* (*e-flux Architecture*, 2020), https://www.e-flux.com/architecture/accumulation/.

Copyright 2024 by the Regents of the University of Minnesota

All rights reserved. No part of this publication may be reproduced, stored in a retrieval system, utilized for purposes of training artificial intelligence technologies, or transmitted, in any form or by any means, electronic, mechanical, photocopying, recording, or otherwise, without the prior written permission of the publisher.

Published by the University of Minnesota Press
111 Third Avenue South, Suite 290
Minneapolis, MN 55401-2520
http://www.upress.umn.edu

ISBN 978-1-5179-1717-3 (hc)
ISBN 978-1-5179-1718-0 (pb)

LC record available at https://lccn.loc.gov/2024041077

Printed on acid-free paper

The University of Minnesota is an equal-opportunity educator and employer.

CONTENTS

Introduction. Age of Resilient, Planetary Urbanization?
Imaginaries and Infrastructures of the Anthropocene 1

1. Miami Imaginaries
Climate Change Ground Zero or
Resilient "Capital of Cool" 43

2. New Houses for a New Century
Urban Futures Scenario Visioning
and Social Infrastructure 81

3. Flows against Flows
The Everglades as Critical Urban Infrastructure 109

4. Beyond Resilience
Islands, Urbicide, and Infrastructural Delinking 135

5. Anthropocene Critical Urban Theory and Practice
Research and Development for a Thawing World 169

Coda: Zone of Transit 199

Acknowledgments 205
Notes 207
Index 271

INTRODUCTION

AGE OF RESILIENT, PLANETARY URBANIZATION?
IMAGINARIES AND INFRASTRUCTURES OF THE ANTHROPOCENE

> Each day it becomes more apparent that the future is . . . a territory being mapped and claimed by contemporary imaginal formations.
> —Natalia Zuluaga, "Point and Pivot"

The Urban Age

When people think of Miami, perhaps they envision afternoons on the beach under the sun, a Spirit Airlines–enabled escape from icy winters in northern cities. Nostalgically, they might also envision *Miami Vice*–esque nighttime drives under luminous neon lights to the sound of Phil Collins's "In the Air Tonight"—or cruising Biscayne Boulevard à la Future's "Mask Off." Equally and more contemporarily, the mind might turn to ventanitas and Elián González, Republicans and culture wars, plastic surgery and cocaine cowboys, or crypto and finance. The urban imaginary of Miami contains all of these and more. But since the 2010s Miami has quickly come to signify something else in the popular and professional imagination as well: climate change ground zero, a modern Atlantis soon to be swallowed by seas, and a front-runner in urban adaptation, a model of proactive resiliency building for planners and governments worldwide. These imaginaries are entwined, shaping and warring with

| 1

one another in turn. But whether imperiled or rising above, what is clear is that Miami is now an Anthropocene city in which the future of twenty-first-century urban life is rapidly unfolding—and on the table as an open question to explore.

Focused on Miami but oriented toward debates and horizons in urban thought and practice more broadly, this book is a critical theoretical and empirical exploration of how diverse actors in the city perceive the problems of climate change and sea-level rise, imagine urban futures, and launch infrastructural design responses. Through these analyses, the book also traces the recalibrations of urban space, nature, and life, both infrastructural and imaginal, at play within such sea-level-rise adaptation designs. Enframing these more specific investigations are a set of broader questions regarding omnipresent imaginaries of the Anthropocene as an age of planetary urbanization and resilience, and the book is also interested in the potential for urban thinkers and practitioners to untether ourselves from this imaginary and investigate the transformations of urban space, life, and futures that are both needed and occurring now.

To begin, a discussion of what exactly is meant by the Anthropocene, as well as its connotations as an urban and resilient age, is in order. Used frequently by earth and social scientists to describe our contemporary historical epoch, the term *Anthropocene* is understood in this book as pointing to the manifold and overlapping social and ecological crises facing liberal industrial societies in the early twenty-first century. The term arrives at a moment when the blip of post–World War II liberal "peace" is over and as soaring economic inequality and debt, hyperexploitation, and algorithmically induced subjective unraveling are now flanked by the collapse of the natural world itself: melting ice caps, rising sea levels, extinction of animals and plants, devastating wildfires, record heat waves, and violent and increasingly unpredictable storms which are no longer surprise events but instead constitute the normal course of things. Industrial activity is now pushing Earth out of the approximately 11,000-year-long Holocene interglacial period of relatively stable climates and rich biodiversity in which modern civilization developed, and into what scientists Will Steffen, Paul Crutzen, and John McNeill describe as "planetary terra incognita."[1] In this moment, the catastrophic social and environmental impacts of industrial liberal regimes have become plain. The codes that once grounded modern thought and action, invented in now long-gone contexts, are being rendered obsolete. Philosopher Giorgio Agamben's Pauline dictum that "passing away is the figure of this world" has become the shared sense of an entire generation.[2]

While emerging from the geological and earth system sciences and frequently understood as a solely environmental matter, over the last decade the Anthropocene has become an important name for this much broader and far-reaching situation—which is as much social, philosophical, and political as it is environmental—and the widespread upheavals in thought and practice it is both inaugurating and being propelled by.[3] As the late philosopher Bernard Stiegler puts it, the Anthropocene is "a test and ordeal of thinking in all its forms (as knowledge of how to do, how to live and how to conceptualize)."[4] Indeed, for many designers, planners, scientists, and critical thinkers, established modes of living, governing, and thinking are now outmoded or ineffective; the frameworks that once grounded modern thought and action, invented in now-long-gone contexts, are no longer sufficient to grasp rapidly mutating realities.[5] Climate change, for example, is widely understood as having rendered modern forms of security and governance, planning and infrastructure, obsolete, and calls for experimentation with new approaches to each are now widespread.[6] Similarly, in critical scholarship, art, and design, statements proliferate to the effect that the Anthropocene "changes everything," marking a rupture point when now-outdated rubrics must be thrown out, imagination and new ideas are needed, and a massive, multidisciplinary process of experimentation with thought, practice, and life itself is seen as imperative.[7] As Holocene codes crumble and critical thinkers explore the epoch's upending of modern ontologies, human/nature binaries, and socio-ecological destruction, they have also emphasized repeatedly that, rather than made by a generic "human," the Anthropocene and its dislocations are produced by historically specific forms of capitalist accumulation and control.[8]

Across diverse analyses, the Anthropocene is regularly characterized in terms of *urbanization*. Whereas some like Jason Moore tie the emergence of the Anthropocene (or in his words, Capitalocene) to exploitations of nature and people as early as the sixteenth century, many other scholars link the epoch's advent to the industrial urban period, identifying capitalist production of urban spaces as both Anthropocene driver and artifact.[9] But perhaps even more ubiquitous than historical and causal analyses are characterizations of the Anthropocene present and *future* as an ever more fully urban age. Here the Anthropocene is routinely linked to postulations of the planet's urbanization, which describes either that the majority of the world's population now lives in cities or the integration of once-distant sites into a planetwide socio-ecological network of processes and technologies.[10] Urban systems are understood as vulnerable to and intertwined with floods, heat, and other socionatural catastrophes.[11] But cities and urban processes are also viewed as climate change solutions.[12] By retrofitting cities with a bricolage of resiliency infrastructures, governments and planners hope that extreme events and threshold crossings will be absorbed and governed, allowing urban spaces to maintain their basic structures and functions.[13] In this way, it is almost always imagined, the inexorable development of urbanization and the Anthropocene—or Urbicene, as Erik Swyngedouw has named it—will remain necessarily companion processes.[14] In fact, for geologist Jan Zalasiewicz, cities will be one of humanity's longest-lasting traces on the planet, future fossils whose imprint will remain long after humans.[15] With the urban envisioned as the inevitable form of the twenty-first century, it seems the only question mark is whether urban spaces and processes will be more or less resilient or equitable, smart or inclusive.

But what mutations of urban space and infrastructure, government and imagination, are being produced as diverse actors attempt to adapt cities to perceived Anthropocenic upheavals? Through what sites and spaces are contemporary adaptation responses to present upheavals—epistemological, social, environmental—being enacted, and what future trajectories do they open and close? Are these destined to be always enrolled as a means of augmenting the resilience of existing urban systems, and is urban resilience the final form of urban responses to climate change? Will the seemingly inexorable development of urbanization and the Anthropocene remain necessarily companion processes, and is planetary urbanization in fact the necessary telos and spatial limit of life in the epoch? Will and should the urban as we know it actually survive the upending impacts of climate change or human responses? Or, if the Anthropocene truly does shatter long-held ecological, epistemological, and structural certainties, could it be that even the contemporary maps of the urban condition we now hold are already becoming obsolete?

Anthropocene Critical Urban Theory/Practice

In series of remarkable articles and books published over the last decade, urban theorist Neil Brenner has issued a powerful challenge to urban scholars. Instead of using outmoded, nineteenth- or twentieth-century spatial concepts to understand twenty-first-century transformations, Brenner calls for us to "explode our inherited assumptions regarding the morphologies, territorializations and sociospatial dynamics of the urban condition" and open our eyes to the mutations in urban form and process produced by contemporary global capitalism.[16] For too long, Brenner argues, urban thinkers have been utilizing spatial concepts and categories invented in long-gone historical moments to understand current realities. Understanding contemporary urban transformations, he wagers, requires "a constant reinvention of the framing categories, methods and assumptions of critical urban theory in relation to the rapidly, unevenly mutating geographies of capitalist urbanization, especially in the contemporary era of hyperfinancialized, planetary-scale spatial, institutional and ecological transformation"—that is, "a process of restless, open-ended theoretical experimentation."[17]

Brenner's call for "urban theory for our time" has been extremely generative, leading scholars to productively push beyond spatial assumptions and critically index manifold "new urban spaces" and "vocabularies of urbanization" to capture how the last several decades of capitalist accumulation have produced the urban as a planetary state of existence.[18] For these thinkers, urbanization is not a single thing but a heterogeneous meshwork of processes and forms of capture which themselves are productive of diverse sites and spaces. Pushing beyond urban thinking focused exclusively on "the" city qua urban agglomeration—now an artifact of earlier eras of urbanization—scholars have tracked the explosion of sites being operationalized to serve agglomerations of urban life, including spaces of resource extraction, massive industrial farms, global supply chains, infrastructures of communication and distribution, and the growing expanse of entangled human and nonhuman spaces and processes which together now constitute urbanization's global "operational landscape."[19]

Brenner's "urban theory for our time" call prompting this explosion of work resonates strongly with the understanding of the Anthropocene indicated above, as a moment of rapid, far-reaching social, political, and environmental transformation in which the foundational frameworks and methods of industrial liberal thought and practice are being profoundly upended, and in which deep experimentation is needed in all realms: infrastructural, epistemological, philosophical, and imaginal.[20] And while not often using the term *Anthropocene*—planetary urbanization scholars frequently prefer *Capitalocene*—much of this work traces how urbanization processes are *producing, driving*, and *amplifying* key aspects of the "Human Epoch." Here we might think of Brenner, Roger Keil, and Swarnabh Ghosh's studies of how globally interconnected urban systems are helping produce and circulate emergent infectious diseases.[21] Or we might think of geographer Martín Arboleda's work on the destructive human and environmental impacts of the global mining sector.[22] However, within critical urban scholarship, far less attention has been given to how urban spatial forms and imaginaries—including those recently identified within planetary urbanization work—are themselves being transformed, reconfigured, and recalibrated by climate change and human adaptation responses. That is, many scholars have considered the entwinements between capitalist urbanization processes and the *production* or *perpetuation* of the Anthropocene. But less work has been done to understand the recalibrations, reconfigurations, and transformations of urban government and urban infrastructure, urban imaginaries and urban life that are being generated via attempts to adapt to and govern the *effects* of the Anthropocene—which are, again, social as well as environmental—on existing urban systems, and to build resilience to the climate-change-related crises that capitalist accumulation and urbanization processes produce. But such engagement is needed if urban theory is to be truly contemporary, for as architect and urban scholar Kian Goh asks, "If climate change is the defining challenge of the moment, how is it not indelibly transforming our core thinking of the urban?"[23]

The point here is not that critical studies of urban climate adaptation are lacking. Critical interrogation of urban resilience, for example, abounds. There is a vast multidisciplinary literature devoted to critically analyzing how the shift toward building resilient cities often extends and is intertwined with socioeconomic inequality, racialized exclusion, neoliberalism, and governmentality.[24] Municipal resilience projects have also been extensively criticized by scholars for failing to include marginalized communities or issues like mental health.[25] Focusing on other, broader scales, critical thinkers have fruitfully exposed how the resilience paradigm is a recalibration of existing neoliberal ideology that frequently overlooks social, economic, or racial disparities while naturalizing extreme precarity.[26] However, theorists and practitioners frequently point out, resilience, and urban resilience more specifically, is not one "thing" that is simply applied everywhere with the same results. Instead, as with urbanization itself, it is better understood as a process—resilient urbanization—a *governmental* as well as *political-economic* process that is already producing its own novel spatial forms and imaginaries.[27]

Indeed, as recent work by urban scholars makes clear, urban climate adaptation is already generating its own recalibrations of space, infrastructure, and imaginaries, as well as novel urban sites and processes. These Anthropocene urban processes and forms, like New York's experimental "living infrastructure," which I have analyzed elsewhere, find good company alongside the Rotterdam–New York–Jakarta global-urban climate adaptation network analyzed by Goh.[28] Similarly, these sit alongside rural coastal Bangladesh's preemptively salinized coastal waterways and attendant commercial shrimp aquaculture sites, which geographer Kasia Paprocki—whose powerful work I return to many times throughout this book—shows are being produced via international financial institutions' efforts to adapt the purportedly sea-level-rise-doomed region for more profitable, adaptive ends.[29] As both Paprocki's and Goh's important work on these sites makes clear, urban resiliency projects create their own novel socio-ecological systems, as well as political conflicts over the future use and design of urban space. These can be seen well in the recent bulldozing of New York City's East River Park—a locally beloved waterfront holdout against the new New York of Shake Shacks, Instagram-ready street art, and Billionaires' Row—to make way for the $1.45 billion East Side Coastal Resiliency Project, a 2.4-mile system of floodwalls and floodgates to govern sea rise and flooding. In place of the park's bulldozed social and ecological landscape (trees, plants, soils, and human structures), which is now said to be obsolete, the city is constructing an elevated "world-class" park with "updated" amenities and new tree species adapted to saltwater exposure and extreme storms that doubles as a seawall.[30] Not just individual infrastructures, insofar as such sites of resilient, adaptive management are being devised to help secure critical nodes of planetary urbanization itself—New York as a powerful financial and cultural center, for example—we should additionally understand them and their peers *together*, as constituting an emerging "stack," to repurpose Benjamin Bratton's term, to govern existing urban spaces and systems.[31]

Still, despite the valuable studies emerging to complement what is already a wealth of critical work on urban resilience and planetary urbanization, there nevertheless remains much less focus within urban scholarship on how resilience-building efforts are restructuring urban spaces and processes, and recalibrating the imaginaries, planning frameworks, and modes of government that shape them. And what remains far more fundamentally unquestioned is the assumption that the basic spatial forms of contemporary cities and urban processes will (and should) remain (more or less resiliently) as the Anthropocene progresses. Assumed within so much urban and Anthropocene scholarship is *urban resilience*: the continuing existence and expansion of the urban (all that is needed, presumably, are better ways of organizing or managing it). But if the Anthropocene is indeed a time of deep environmental and epistemological uncertainty in which long-held assumptions and processes are being upended, might it be that even our more recently "inherited cognitive maps" of the planetary urban condition are themselves soon to be out of date, scrambled by climate change and adaptive strategies?[32] After all, concurrent with projections of planetary urbanization as an inevitable condition of the Anthropocene, in coastal cities from Lagos to New Orleans, alternative visions of the urban's *end* are now emerging. Rather than an endless expanse of cities and urbanization processes with seemingly no terminus—the latter destined to be but fodder for ever greater resilience of the former—might the Anthropocene's human and nonhuman dislocations produce other spaces, processes, and imaginaries entirely?

The City at the "End"

These questions are not at the forefront of most critical urban theoretical agendas, but this book argues that they should be. Exploring them, I argue, requires a new research agenda appropriate to the contemporary historical and political moment. In keeping with the passing away of modern frameworks and grounds that the Anthropocene names, such an Anthropocene critical urban theory/practice—as I propose we call it—resists the temptation to rely on inherited spatial concepts—and when appropriate, even newer ones like planetary urbanization or resilience; abandons assumptions as to what urban life in the Anthropocene is, can, or will be; and attends to how twenty-first-century Anthropocenic urban transformations are producing novel spatial and imaginal mutations, including nascent spaces and visions of human and nonhuman life that push beyond currently dominant paradigms such as urban resilience. If urban thinkers want to avoid being "late to the party," as Keil puts it, we need an urban theoretical/practical approach that amplifies and expands on Brenner's "urban theory for our time" challenge—and takes it to its fullest possible conclusions.

Contributing modestly to this orientation and building on the above-mentioned planetary urbanization and critical resilience studies, this book considers these matters via the first critical theoretical and empirical exploration of experimental sea-level-rise adaptations in Greater Miami.[33] Located at the southeastern tip of Florida and today home to 6.1 million people, the sun-scorched, densely developed metropolitan area is wedged between the urban-expansion-constraining Everglades and the Atlantic Ocean (Figure 1).[34] In the late nineteenth and early twentieth centuries, Black and white laborers from the American South as well as Black immigrants from the Bahamas carved the city out of drained wetlands and mangrove swamps. Today Miami is a majority-Hispanic city shaped culturally, politically, and architecturally by multiple waves of immigration. Home to some of the richest and poorest neighborhoods in the United

Figure 1. Florida, as seen from the International Space Station, in a photograph captured by NASA astronauts. The Miami metropolitan area is the bright stretch of light on the peninsula's southeastern tip, flanked to the west by the Everglades, a regional wetlands ecosystem visible in the vast concentration of darkness to Miami's left, and the Atlantic Ocean, to the east and south. Copyright: NASA Earth Observatory, October 13, 2014.

States and built on Jim Crow legacies of racial segregation and political marginalization, Miami is an urban agglomeration of extreme socioeconomic polarization.[35] Often described as the financial capital of Latin America due to the city's role as an economic connector or "entrepôt" linking the global North and South via trade and financial flows, Miami has an economy that is largely dependent on tourism, investment, and real estate development.[36] Foreign capital and an endlessly evolving imaginary of urban tropical paradise drive the city's hypersegregated real estate market; notably, the city has been and remains a hub for money laundering for both international investors and drug cartels alike. And, as will be discussed in depth in chapter 1, post–Covid 19 Miami has also become a key node for finance, tech, and cryptocurrency firms and investors.

In recent years Miami's political-economic landscape has intersected with climate change effects—both actual and projected—threatening the city's financial as well as physical foundation itself and quickly transforming the city into what one might, with a nod to Edward Soja, call the very capital of the Anthropocene. Built on porous limestone made of marine deposits accumulated from the remains of sea creatures over millions of years when the area was underwater, the city is experiencing early and increasingly frequent sea-rise flooding not only along coasts but also from water coming up through its permeable foundation. During annual King Tides, sunny-day flooding sends water gushing out of some street drains and into roads and sidewalks even on non-rainy days multiple times per year, events that are increasingly normalized into the fabric of the city in the form of LED flood alert signs and King Tide awareness posters.[37] Along with floods, the city is also experiencing saltwater intrusion in drinking water supplies and beginning septic tank failure from rising water tables.[38] In addition to being one of the lowest-lying coastal metropolitan areas in the United States, Miami is also classified as a sea-level-rise "hot spot." According to scientists Arnoldo Valle-Levinson, Andrea Dutton, and Jon Martin, this area of accelerated rise stretches from the southeastern tip of Florida up to Cape Hatteras, North Carolina, and, between 2011 and 2015, saw sea levels rise over three times faster than the global mean.[39] The densely developed subtropical region also faces extreme heat, intense tropical storms, and severe, flooding-inducing rainfall events.[40]

Whereas sea-level rise still appears as a future matter for many cities, these unique and combined geographical factors have led to Miami's frequent portrayal as sea-rise ground zero in the United States. Indeed, Florida is used frequently as an example of climate change impacts in the International Panel on Climate Change's 2022 report,[41] which states that Florida's socio-ecological landscape has already been radically transformed by climate change. Thus do the images of sunny beaches or spring break for which Miami is known now coexist with a new set of iconic pictures capturing—and amplifying perception of—these transformations, and particularly the city's floods, such as Miamians driving through submerged intersections on their way to work or Miami-based Russian photographer Anastasia Samoylova's color-saturated photos of flooded apartment complexes and roads (Figure 2 and cover). Such images capture the transformations already underway in Miami, but are also understood and forwarded as a preview of the city's coming aquatic future, with sea-level rise projected to increase rapidly over coming decades: 10–17 inches by 2040; 21–54 inches by 2070; and 40–136 inches by 2120.[42] As sea levels rise, high-tide

Figure 2. *Staircase at King Tide*, from Anastasia Samoylova's *Floridas* series. Copyright: Anastasia Samoylova, 2019.

flooding is projected to increase to 380 events per year by 2045, followed by "chronic flooding approaching permanent inundation" in some areas.[43]

Such projections are also widely viewed as a warning of Miami's political, economic, and infrastructural vulnerability. Miami is ranked number one worldwide in terms of real estate and infrastructural assets exposed to sea-level-rise-related flooding, and will have $3.5 trillion vulnerable to coastal flooding by 2070.[44] According to a 2020 Resources for the Future report, Miami "faces the largest risk of any major coastal city in the world" with $400 billion in assets at risk.[45] A 2020 Urban Land Institute report projected the loss of over $4.2 billion in property to daily tidal flooding by 2040, followed by the potential loss of $ $53.6 billion worth of property by 2070.[46] Along with financial and real estate risk, sea rise is seen as a key threat to the city's infrastructures, with hurricanes and rising seas threatening to cripple roads and energy networks. Here, within media and government reports, a new Anthropocenic map of Miami emerges, one whose key landmarks are imperiled critical infrastructures and real estate. Downtown Miami, with its luxury waterfront development and cruise ship port—the world's largest—faces potential submergence with six feet of sea rise. But more immediately, salt water is already contaminating urban drinking water systems; with additional sea-level rise, experts warn that it may contaminate tap water and render sewage systems nonfunctional.[47]

Reimagined and reproblematized as sea-rise ground zero, Miami is now heralded by politicians, engineers, and planners as a climate change first responder and an observatory for the issues other coastal cities worldwide will soon face. Whereas cities in the Netherlands use dikes to mitigate flooding or Manhattan might be secured using massive Bjarke Ingels–designed sea barriers, Miami's permeable foundation means flooding can advance even with walls, producing a unique challenge for adaptation planning and design. To ward off mediatized "Goodbye

Miami" projections of the city as the first urban casualty of climate change this century, politicians and designers, scientists and engineers have enrolled the city as a "living laboratory" for experimental resiliency infrastructures.[48] Whereas former Florida governor Rick Scott publicly denied climate change—and issued an unofficial ban on use of phrases like "climate change," "global warming," and "sea level rise" in government documents—in recent years the compound threats climate change poses to South Florida's infrastructure and economy have become key governmental issues, especially at the city and regional level.[49] In 2006 Miami-Dade County created the Climate Change Advisory Task Force, which made some of the Miami area's first official recommendations for sea-rise adaptation and proposed Lidar mapping of critical infrastructure elevations vis-à-vis sea-level rise. The City of Miami, the City of Miami Beach, and Miami-Dade County have each created chief resilience officer positions to oversee municipal resilience planning. In 2019 their Greater Miami and the Beaches partnership released *Resilient 305*, a strategy plan for fielding increasingly frequent shocks and stressors, from sea rise and hurricanes to socioeconomic inequality and traffic jams.[50] In 2023 the Biden-Harris administration announced the formation of the South Florida Climate Resilience Tech Hub, a consortium led by Miami-Dade County, opening the region to $75 million in federal funding for resilience building and promising that resilience efforts will continue in the near future.[51] The ongoing subject of many university architecture and design studio seminars, including Harvard Graduate School of Design, and consultations with Dutch engineers, virtual reality sea-rise visioning, and future scenario workshops, the city has, for diverse actors and interests, in general become a critical site for reimagining and redesigning urban futures.[52] Most often, the central concern of these efforts is how to ensure the resilience of Miami's current social, economic, environmental, and infrastructural systems—at least in the near future.

Urban Resilience

This new thinking of the city has brought Miami to the center of an ongoing resilience problematization and practice of urban government in the age of climate change. Since at least 2012, when Hurricane Sandy struck New York City, resilience has supplanted sustainability in many quarters as the dominant governmental methodology and discourse under which a host of technologies and designs are being stitched together in hopes of managing urban systems. Developed in the 1970s by ecologist C. S. Holling as a mode of managing the adaptive ecosystems covered in his research, *resilience* is defined as "the capacity of a system to absorb disturbance and still retain its basic function and structure."[53] In contrast to what are now seen as outdated modern modes of management that sought to maintain a single stability state via top-down techniques of control and regulation, resilience is heralded as a flexible, postmodern form of management that seeks to create and define safe operating spaces able to manage, rather than eliminate, interruptions and crises. With twentieth-century urban governance approaches seen as cause of, and insufficient to manage, present and future disasters, resilience has been picked up from ecology and development thinking and celebrated by city planners, designers, and municipal and funding bodies alike as the city's salvation and best hope of managing the systems of contemporary urban life amid climate change, rising seas, and infrastructure vulnerability.

Within ecology, where the concept emerged through the critical problematization of existing environmental management practices, resilience is seen by proponents as a corrective to what are now viewed as "pathological" "command and control" management approaches focused on maintaining isolated systems in equilibrium states by preventing disruption.[54] In place of such methods, systems ecologists devised resilience as what they saw as a more realistic adaptive management approach using experimental techniques to absorb and attenuate shocks in nonequilibrium systems by developing "the capacity of a system to absorb disturbance and reorganize while undergoing change so as to still retain essentially the same function, structure and feedbacks, and therefore identity."[55] While the early formulation of resilience was derived from empirical work on adaptive ecosystem management, by the 1990s resilience thinkers had expanded the concept to the governing of the social, approaching human institutions, organizations, and agents as part of complex socio-ecological systems.[56] This work was extended by Holling and colleagues in the Resilience Alliance and analyzed past societal collapses, institutional and economic management, social capital, and sustainability.[57]

Recent years have seen the widespread translation of resilience into urban governance, planning, and design. From Miami to New York, Lagos to New Orleans, city planning commissions, art exhibitions, and newspapers all repeat similar, resilience-inflected catchphrases, creating an echo chamber in which resilience now rings as the best and only refrain. Based on a view of crisis as inevitable, cities as coupled social-ecological-technical systems, and techniques for absorbing or withstanding turbulence, making cities resilient is widely seen as the appropriate solution to the Anthropocene as it has (for now) been defined. Here definitions echo those of ecology but are shifted into an urban infrastructural context. Instead of sustainability or the maintenance of systems in an existing steady state, resilience approaches seek to build adaptive capacities for maintaining a system's essential parameters and identity through the allowance and harnessing of turbulence. Depicted variously as the capacity to "ride out" or "bounce forward" from extreme events, resilience is not, as proponents often depict it, an apolitical or ahistorical techno-managerial approach. Rather, as the critical work outlined in the above section has shown, resilience is a *mode of governing* urban populations and environments in the Anthropocene. It is also a *process of urbanization* that combines adaptive management with anticipatory critical infrastructure protection to build cities' capacities to field, absorb, and govern crises so as to maintain liberal urban life and economies. As such, sought across manifold resilience efforts are adaptive eco-cybernetic urban systems able to administer socio-ecological shocks and stresses while maintaining their essential parameters. These Anthropocenic infrastructures and urbanization processes are being added to the twenty-first-century urbanization nexus that Brenner describes as the "evolving, mutually recursive relations between agglomeration processes and their operational landscapes."[58] And, given their critical function of *administering* and *securing* the continuation of existing urban agglomerations and operational landscapes alike, resiliency infrastructures are, perhaps, deserving of their own distinct conceptual category within this nexus.

Faced with urban futures promised to be hot, crowded, and tremendously polarized—a planet of increasingly tense cities where natural disasters and new forms of urban insurgencies will intertwine to produce previously unseen effects—governments and planners are using cities and urbanized spaces as laboratories for trialing infrastructures able to administer political, economic, and technological systems amid climate change. Here, abandoning long-held planning approaches and embracing experimentation are seen by practitioners as key to urban resilience.[59] Urban laboratories (such as Miami is now considered) are narrated as offering opportunities for quickly testing local, situated resiliency solutions within and in response to local crises, environments, histories, and metabolisms. Following resilience's experimental methodology, universities and think tanks, architects and city commissions are transforming cities or parts of cities—blocks, sidewalks, neighborhoods, communities—into design test runs (or, to use Orit Halpern's phrase, what we might call "urban test beds").[60] In this situated experimentation, the lab is neither a pristine or sterile space—the modern laboratory—nor a single dedicated space but rather the urban environment itself. There, resilience-building efforts intersect with local conditions and contexts, producing unique spatial and imaginal effects, as well as definitions of urban life, in the process.[61]

Problem-Based Urban Studies

To explore these emergent spaces and imaginaries, one cannot rely on preexisting rubrics, tropes, or ideological assumptions. Instead, thinking the city in the Anthropocene requires that we set aside neat or coherent frames already laid out for us and instead explore how particular logics such as resilience intersect with local contexts. This is because, as many scholars emphasize, urban resilience is itself not a preexisting "thing" but rather a context-specific, ad hoc assemblage of heterogeneous elements that manifests differently across place and time.[62] Rather than a preexisting logic or inevitable paradigm, urban resilience should be analyzed critically as what Foucault called a dispositif, an ad hoc arrangement of techniques—discourses, practices, architectural forms, regulations, laws, knowledges, technologies, and designs—brought together to govern a crisis, in this case climate change, as well as climate change imaginaries, and their potential effects.[63] Thus, rather than assuming one knows in advance, to analyze urban resilience in Miami or any other city we should ask: How do diverse urban actors perceive the problems of rising seas, imagine urban futures, and enact situated responses? What novel urban infrastructures, imaginaries, and definitions of human and nonhuman life are generated therein, and how are these shaped by specific environmental, social, political, and infrastructural contexts and legacies? What political possibilities are opened and closed in these efforts, and what new spaces of experimentation can be developed?

In what follows I explore these questions using a problem-based approach.[64] Building on Clive Barnett and Gary Bridge's call for a "problem responsiveness"–based urban studies, as well as Foucault's earlier work on "problematization," I analyze climate resilience experimentation in Miami as a context-specific arrangement of techniques and imaginaries at diverse sites and scales that reimagine and often re-create urban government, infrastructure, and life.[65] This problem-based approach is especially suited to Anthropocene critical urban theory/practice, and so a few words on it here are useful. For Foucault, problem-based analysis does not "problematize" an object of critique in order to unmask or reveal what it elides. Instead problem-based studies focus on crisis moments, which may be single events but more often in Foucault's thought are moments of upheaval and shifts in practices or strategies when certainties become uncertain, difficulties are encountered, and thought must reflect anew.[66] Within such moments, problem-based analysis tracks the shaping of phenomena into coherent problems posed in particular terms and the creation of specific parameters within which responses can be posed, while indexing the framing of specific responses as appropriate (and others as inappropriate).[67] Using this framework allows me to track how solutions arise from specific ways of framing problems and how, rather than natural or inevitable—as they are often presented—these solutions themselves are assembled.[68] While a problematization may sometimes emanate from a single actor, in Foucault's work problematization occurs through a confluence of diverse sources: discourses, representations, concrete practices, laws, architectural designs, and so forth. Rather than a linear history, which shows the progressive development of something that already exists—"resilience," "sexuality," and the like—problem-based analysis pays attention to discontinuous assemblages of techniques brought together in space and time on the basis of specific problematizations and urgencies, how they are stylized, stitched together, and made coherent.

The element of uncertainty and reactivity is crucial to the problematization method. As Clive Barnett has argued, the convention of critical theorists to focus on demonstrating the constructedness of governmental regimes—thus "puncturing the appearance of naturalness or inevitability"—downplays the element of problem responsiveness that Foucault himself saw as essential to such regimes.[69] Indeed, in "The Confession of the Flesh," where Foucault most clearly defines the notion of the dispositif, he indicates that the concept must be understood not only in terms of being an ad hoc arrangement of heterogeneous elements but equally with regard to its "genesis" as a strategic response to a perceived problem or "urgent need."[70] More broadly, Colin Koopman argues, it is Foucault's rigorous philosophical and genealogical attention to the problems grounding the present that constitutes his work's critical force and which yielded powerful concepts as dispositif or biopower in the first place.[71] In his later work, Foucault articulates this problem-based method as itself an alternative practice of critique, a "critical interrogation of the present" that, along with identifying "the limits that are imposed on us," is an "attitude" and "ethos" that explores "the possibility of going beyond them."[72]

Historical examples of problems analyzed by Foucault include crime and delinquency, sexuality, and mental illness.[73] Since such phenomena may have existed previously but were not always considered problems, the question for problem-based analyses is how certain phenomena come to be perceived as problems at a given historical moment, how specific responses are crafted as solutions to them, and what kinds of subjectivities are generated therein.[74] But just as important, this book will emphasize, is an attentiveness to emergent problematizations through which even recently defined frameworks (such as urban resilience) themselves are already being questioned or potentially supplanted.

This method is a central part of what I am calling Anthropocene critical urban theory/practice. In the Anthropocene—itself a "problem space" in the broader temporal sense used by Paul Rabinow—sedimented ways of thinking and acting are coming undone (or being dismantled, recalibrated, etc.).[75] Rather than rehashing preexisting tropes that are the same everywhere, or analytical frames imported from the past, analyzing contemporary urban resilience requires attending to how its projects are designed precisely amid and in response *to* this dissolution of inherited tropes, urban-planning approaches, and epistemologies. As Foucault emphasized, problematizations occur and must be analyzed in terms of specific contexts.[76] And so, as geographer Bruce Braun puts it, what we need to understand is the way in which established contexts and logics (political, economic, imaginal, etc.) critically "encounter" Miami's unique social, political, economic, environmental, and geological contours and legacies.[77] These include, for example, geology. The limestone rock on which the city is built is already producing a fundamentally distinct manifestation and experience—and mediatized spectacle—of sea-level-rise flooding, in the form of saltwater gurgling up through street drains and underground foundations (rather than only from coasts). As we will see in chapter 3, this geological context is also producing its own unique urban risks as well as corresponding definitions of infrastructure. But along with geology, one also has to understand how the adaptation experiments analyzed in this book intersect with and are shaped by the city's existing landscape of extreme economic inequality and real estate and imaginative speculation. As Kevin Grove, Savannah Cox, and Allain Barnett argue, for example, municipal resilience building in Miami is being shaped through the city's history of segregation and political marginalization and is producing its own new forms of political, economic, and racialized injustice.[78] And, in reverse, it is also important to consider how these local contexts intersect with broader epistemological shifts and imaginaries, such as widespread turn within fields such as urban planning, design, and ecology toward understanding cities as complex more-than-human systems. Whether negative or positive, the point of a problem-based approach is that instead of simply seeing what we already see or want to see, attuning instead to these conjunctures—however contradictory they may be—forces us to actually *think*, freely *imagine*, and *see* emergent novelties, including, in some cases, even those we may not wish to see.

By emphasizing novelty and emergent urban forms—which frequently outstrip urban scholars' existing concepts, imaginaries, and vocabularies—it is not my intention to suggest that the urban resilience infrastructures or imaginaries I analyze arise out of nowhere. Nor am I suggesting that they represent a clean, decisive break with urban infrastructure or imaginaries past. Rather, my interest is in how these are being remodulated, reimagined, or defined in response to historically specific problematizations of urban government, nature, and life. Urbanization has long been, after all, an ongoing *process* of critique and creative destruction, in which each style of urbanization—often launched in response to perceived crises—is influenced by those that precede it.[79] As I show in several places, such as in chapter 3's discussion of critiques of urban planning that have existed since the 1960s, or in chapter 2's discussion of social infrastructure, urban resilience approaches draw on and rework elements and energies that already exist, albeit in many cases by assembling them together in new ways and toward different ends (in this case, toward managing impacts of climate change on urban systems). And certainly, even more recent innovations in urban resilience infrastructure or imagination coexist with other urban forms and planning approaches (command-and-control engineering projects, master plans that block out water, etc.). As Foucault always emphasized, it is never a question of marking a clean historical break, but instead of tracing out new contours and problematizations of thought and practice as they are brought together in distinct historical moments and places. As these assemblages cohere, urban forms, modes of government, infrastructure, and imagination become possible that, historically, were not. Thus, what I am interested in tracking throughout the book are these historically specific problematizations, and the recalibrations, reconfigurations, and transformations of urban infrastructure and imagination that are brought to appear as responses to them. While these may draw on existing discourses and planning practices, it is important to analyze how they are also distinct from them. And rather than assume urbanization as an eternal, natural, and unceasingly resilient process, a dialectic destined to morph and evolve interminably as the Anthropocene progresses, the book's final wager is to ask whether the epoch might instead inaugurate possible *breaks* from the urbanization process itself.

Urban Infrastructure and Imagination

This book explores these matters as they are playing out within Miami's unique social, political, economic, and geological context. Drawing on site and participant observation, visual and discourse analysis of media and policy documents, and stakeholder interviews conducted as part of a three-year multi-sited research program funded by the Urban Studies Foundation, I critically analyze the social, political, and environmental dimensions of a series of imagined and actually existing adaptation efforts underway across the city. Specifically, I analyze climate adaptation planning and design experiments being designed and imagined by diverse actors at different scales and from diverse socioeconomic backgrounds, including municipal governments, engineers, low-income communities, nonprofits, scientists, academics, water managers, elected officials, and others. Through these, I explore remodulations of urban infrastructures and imaginaries emerging within Miami's unique historical-geological conjuncture. Throughout I am also attentive to the broader conceptual and practical questions these explorations raise for urban thinkers and dwellers, and the political potentials they may open via what I call Anthropocene critical urban theory/practice. Moving between theoretical and empirical analysis, in analyzing these cases, I forefront recalibrations they are making within two distinct but related thematics: imagination and infrastructure.

Infrastructure

Infrastructure has become a key political question of the Anthropocene, and the last decade has seen the flourishing of critical infrastructure studies.[80] Investigating the spaces, landscapes, and geographies of infrastructure both past and present, work in this wide-ranging and multidisciplinary field has opened the infrastructural "black box" by mapping commodity and logistics chains, telling stories of "stuff," or following the "lives" of things from papayas to sneakers.[81] Other studies take the form of ethnographies, sensory accounts, or more philosophical meditations on the meaning, nature, temporality, and political conflicts of infrastructure itself.[82] One of the most generative insights of critical infrastructure studies is that the built environment is not neutral but political, and diverse authors have traced the ways in which politics and power are infrastructural, both product and means of governing the life of populations and environments.[83] Many of these studies extend Foucault's perspective that government does not reside exclusively or even primarily in governments but consists rather in ad hoc assemblages of technologies and designs, architectures and infrastructures that make up the built environment.[84] As anthropologist Brian Larkin writes, "liberalism is a form of government that disavows itself, seeking to organize populations and territories through technological domains that seem far removed from formal political institutions."[85] Put more explicitly, infrastructural spaces constitute what Keller Easterling calls "extrastatecraft," "the secret weapon of the most powerful people in the world," the "unstated" "content manager dictating the rules of the game in the urban milieu."[86] In these studies, infrastructures are understood to be powerful political devices because often they do not appear to be doing anything and even appear "natural" or simply "there."

This book builds on this work, exploring the political imaginaries and futures forwarded in emergent Anthropocene urban infrastructure. However, my interest is less to bring infrastructure to the fore and primarily in how—within discourse, media, planning documents and exercises, scientific and other expert statements—infrastructure itself is being reproblematized, redefined, and reconstructed in the epoch. Here I agree with anthropologist Yannis Kallianos's observation that "one can deploy infrastructures as critical kaleidoscopic devices through which to explore the current crisis and its implications for political processes and urban governance," and throughout the book I also trace the crisis or dislocations of nineteenth- and twentieth-century urban government, space, and imagination prompted by the Anthropocene.[87]

Throughout the nineteenth and twentieth centuries, infrastructure was, as Swati Chattopadhyay puts it, the heart of modern urban thought.[88] Critical to liberal capitalist urbanization and control, infrastructure was key to the reshaping of cities into circulatory spaces available for surveillance and unimpeded movement of commodities and workers.[89] Just as natural environments were governed and covered with hard, clean, or metal structures of electrical power, dams, and highways, so too were wild populations disciplined with pipelines and automated grids, laid out by companies to neutralize worker power to strike, suburbs to disaggregate urban and workplace concentration, or the manifold other infrastructures tying them to wage labor and supply chains.[90] Highways, irrigation, and electricity have been central to modern development, and "the provision of infrastructures," writes Larkin, "is so intimately caught up with the sense of shaping modern society and realizing the future . . . [that] possession of electricity, railways, and running water came to define civilization itself."[91] As we will see throughout this book, this relationship between urban life and infrastructure has been central to Miami's own development as well; for example, N. D. B. Connolly argues that state "infrastructural power" over real estate—in the form of Federal Housing Administration loans, slum clearance, and public housing construction, among other matters—played an essential role in the maintenance of social peace, economic growth, and racialized exclusion in South Florida in the postwar period.[92] Similarly, we might consider how the massive engineering effort to drain and compartmentalize the Everglades in the nineteenth and twentieth centuries through the construction of a colossal series of water management infrastructures was instrumental to the development of Miami itself.

But in the opening decades of the twenty-first century, infrastructure—and more specifically, what governments call "critical infrastructure"—is taking on new meanings and forms of significance. Since September 11, 2001, there has been a rise in thinking around securing critical infrastructure, which typically means human-made technical structures such as bridges, roads, electrical grids, smart grids, anti-terror barriers, and surveillance: vital lifelines for industrial liberal urban orders which are now the object of extensive protection measures.[93] But as increasingly frequent storms and floods knock out brittle, vulnerable electrical grids, subway lines, roads, and communication networks, and the threats climate change poses to existing political and economic systems become apparent, along with embedding traditional technical infrastructures into urban environments, urban planners and scientists are experimenting with reinventing infrastructure itself. As I will discuss in this book, here the Anthropocene has seen a broader shift toward the idea of humans and nature as infrastructure in cities, and cities as integrated social-ecological-technological systems of infrastructural systems. All of these are new definitions of infrastructure that bring forth unique technical challenges as well as philosophical and political questions.

Numerous books and articles focus on the need to rethink and redefine urban infrastructure to overcome modern nature/human binaries and adapt cities for the Anthropocene. These works have the virtue of explicitly joining the urban and the Anthropocene, and elucidate how the latter poses important questions and challenges to the former. However, too often, applied resilience approaches to the urban Anthropocene view urban climate adaptation as a merely technical matter and remain uncritical of their own discourse's relationship to transformations in contemporary government. In other cases, critical approaches target urban resilience's lack of inclusion or injustices but remain uncritical of how the turn toward entangled urban–natural systems and calls for rethinking infrastructure are themselves political problematizations being produced in order to secure existing systems. In contrast, this book situates discourse regarding the need to rethink urban infrastructure as itself part of a new, historically specific problematization of urban governance, one in which the Anthropocene's perceived problems and appropriate responses are being crafted and disseminated in specific ways via media, discourse, and infrastructural design. This book critically analyzes these new problematizations of infrastructure, tracking the futures and imaginaries they produce and the political possibilities they foreclose and open.

Imagination and Spatial Imaginaries

The book also explores the distinct but related spatial imaginaries—and with them, possibilities for human agency and imagination—that shape and are shaped by infrastructures, plans, exercises, discourses, and media circulating in Miami vis-à-vis urban resilience and urban futures in the Anthropocene. Here, building on a large body of scholarship concerned with spatial imaginaries, I distinguish between two concepts often used interchangeably in casual conversation: imagination and the imaginary. *Imagination*, on one hand, refers to capacities exercised by individual subjects. Not simply a capacity to fantasize, to envision that which is not real, in philosopher Chiara Bottici's words, imagination is

> the capacity to produce images in the most general sense of the term, independently of whether or not what they represent actually exists; in this view, imagination includes the capacity to represent what does not exist, but it is also not limited to this. It is a much more radical view, in that it includes the production of images of both existing and nonexisting objects. In terms used by Cornelius Castoriadis . . . it is the radical faculty to produce images in the sense of Bilder, without which there would not be any thought at all.[94]

Whereas imagination understood in this way is, as political theorist Julian Reid puts it, "a capacity of the self,"[95] the concept of the *imaginary* has in turn been employed by diverse thinkers to refer to a *social* phenomenon and *context*—the broader images and visions that run through and shape societies differently at different moments.[96]

Importantly, imaginaries are both created and internalized by individuals but are at the same time larger than and beyond them, circulating across media, discourse, art, science, governmental documents and plans, and so on, shaping and limiting possibilities for individual imagination. Building on this notion, many geographers, urban scholars, and spatial thinkers have emphasized the *spatial* and *geographical* nature of imaginaries, and how they are tied to and produced through space and place.[97] Accenting this element, Simin Davoudi gives the following definition of spatial imaginaries:

> Spatial imaginaries are deeply held, collective understandings of socio-spatial relations that are performed by, give sense to, make possible and change collective socio-spatial structures. They are produced through political struggles over the conceptions, perceptions and lived experiences of place. They are circulated and propagated through images, stories, texts, data, algorithms and performances. They are infused by relations of power in which contestation and resistance are ever-present.[98]

The concept of the spatial or geographical imaginary has provided a valuable lens through which diverse thinkers have analyzed the shifting imaginaries through which groups or individuals visualize and model space and place, how modes of sight and visual representation exert powerful social and political effects, and how imaginaries of particular localities are tied up in power relations and productive of differently scaled models and practices.[99] Spatial imaginaries here are understood not as outside to but an integral part of the real, produced through, and constitutive of, social and technological practices. Spatial imaginaries shape subjectivity, power, and the ways humans interact with space, organizing and delimiting how they can imagine and enact the political and demarcating acceptable places of nature and society.[100] Techniques of order as much as of transformation, spatial imaginaries delimit not only what can be known about a particular space but also what is and is not possible in it: "the condition of both the known world and the horizons of possible worlds," as geographer Stephen Daniels puts it.[101]

Material as much as mental, geographical imaginaries are produced, reproduced, and contested in myriad ways through technologies of knowledge and data visualization, media, images, theory, conversation, or social contestation.[102] They give birth to specific practices, perceived urgencies, experiences, and human–nature relations. They condition the sense of what should be, what can be, and, importantly, what will, naturally, be. Perhaps most importantly, they produce certain forms of human subjectivity, while limiting the emergence of others, and narrate spaces, places, and forms of life as in need of certain forms of intervention. In this sense, spatial imaginaries and associated technologies are critical parts of the problematization process described in the preceding section. Writing of mapping and contemporary Big Data–driven urban geographical imaginaries, geographer Taylor Shelton aptly emphasizes how spatial imaginaries do not simply represent existing urban phenomena, but rather "construct new understandings of the urban."[103] He continues: "It is less important that these analyses be correct or comprehensive, per se, but rather that through such analyses, these technologies (and the systems of human knowledge organized around them) enable a particular way of seeing social processes and understanding them as problems to be (and that *can* be) solved."[104] Shelton concludes: "The act of mapping a given urban problem doesn't necessarily produce a meaningful understanding of the problem, though it does, however, help to produce the problem itself, and particular understandings of space underlying it."[105] In highlighting the role of spatial imaginaries in problematizing specifically *urban* space, Shelton also points to the long imbrication of imagination, visualization, and urban problems and futures. As Davoudi points out, "imagining urban futures is as old as Plato's description of the ideal city state in *The Republic*."[106] And far from "merely" aesthetic concerns—a Platonic "less real" to the concrete real—visualization and spatial imaginaries have long been central to urban planning and design, a key means through which urban problems have been defined and specific solutions portrayed as necessary.[107]

Critical work on geographical imaginaries has focused extensively on the visual forms that shaped modern liberal regimes and urban spaces, including those of colonialism, imperialism, militarized visions of cities as spaces of control and theaters of experimentation, the world as picture and object, nature/society binaries, sociotechnical future imaginaries, and twentieth-century capitalist urbanization and globalization.[108] Importantly, these imaginaries do not come from a single source; instead, much like Foucault's concept of the dispositif, they flow across and emanate from urban planning and policymaking documents, data visualizations and geospatial models, government, media, and expert discourse, academic scholarship, architecture, science, art, literature, and so on. Through repetition and circulation across these and other channels, imaginaries become solidified and influential, and critical work on spatial imaginaries has also attended to the visualization technologies that have produced urban spatial imaginaries and urban space itself, as well as knowledge about geography, space, and place.

In recent decades, massive social, economic, and environmental transformations have reshaped liberal societies and urban space, and these have been intertwined with and productive of their own novel spatial imaginaries, such as those of climate change.[109] Here, apocalyptic imaginaries loom large, and have become ubiquitous across literature, film, and popular discourse.[110] But climate crisis imaginaries extend beyond movies and climate fiction to exert very material, political impacts. For Swyngedouw, the imaginary of climate change, "sutured by millennial fears, sustained by an apocalyptic rhetoric and representational tactics, and by a series of performative gestures signaling an over-whelming, mind-boggling danger, one that threatens to undermine the very coordinates of our everyday lives and routines, and may shake up the foundations of all we took and take for granted,"[111] renders diverse adaptation and mitigation initiatives unquestionable and creates a postpolitical consensus in which political conflict is replaced by a regime of technical neoliberal management.[112] The sociotechnical work of constructing climate imaginaries is critical, and Bratton argues that even the idea of global climate change itself would not exist without planetary computation technologies for simulating, sensing, and projecting.[113] Relatedly, technologies for imagining future climate change impacts, such as sea-level-rise charts, interactive maps, or virtual reality tools, at once sketch out new imaginaries of urban futures and enable manifold adaptation interventions—"the climate change of your desires," as Paprocki puts it[114]—a relationship between imaginaries and urban climate resilience that will be discussed later in this book.

Here visions of governing urban space in the age of climate change resonate and often intersect with Big Data and smart city imaginaries, data-driven sociotechnical imaginaries of the urban which are themselves productive of new forms of knowledge and management.[115] However, as Kathryn Yusoff and Jennifer Gabrys point out, apocalyptic imaginaries can also function as a means of critiquing dominant ideologies and realities, providing a necessary beginning point for transforming existing conditions so as to be otherwise.[116] And, notably, climate change imaginaries are themselves multiple; for David Levy and André Spicer, a climate imaginary is "a shared socio-semiotic system of cultural values and meanings associated with climate change and appropriate economic responses," a useful definition which reminds us that there is not one but many potential climate imaginaries, each containing its own understanding of what climate change is, its relative importance, and appropriate responses.[117] In this sense, Edward Said's famous line that "the struggle over geography is not only about soldiers and cannons, but also about ideas, about forms, about images and imaginings"[118] will be especially true in the age of climate change.

Building on this work, in this book I explore the imaginaries operative in and produced by experimental adaptation designs in Miami, and how these imaginaries recalibrate received understandings of urban space, infrastructure, and life. Here, as sociologist Hillary Angelo has done so well with the "city lens" and the notion of urbanized nature, I seek to contribute to analysis not only of the spatial dimensions of contemporary urbanization processes but also the ways of seeing and imaginaries emerging from and driving them.[119] Equally, I consider how these imaginaries open and close our capacity to envision other futures and styles of life than those already present. Finally, through these analyses, I also raise questions regarding embedded imaginaries and assumptions that condition and infuse urban thought and practice in the Anthropocene. In doing so, I seek to advance understanding of urban resilience's own specific imaginary broadly speaking, by exploring some of the perhaps unexpected venues and technologies in which it is sketched out, while also advancing understanding of contemporary mutations within the unique spatial imaginaries that have long undergirded Miami specifically.

Imagination and imaginaries are especially important in Miami, which is and has been since its inception a city of image. "Fantasy," T. D. Allman has written, "has the force of history in Miami . . . in the same literal, matter-of fact way you'd say steel has the force of history in Youngstown, Ohio."[120] Much of this is associated with boosterist manufactured imaginaries of year-round sunshine and orange blossoms designed to fuel real estate investment, or, alternatively, erotic imaginaries of Miami as a place of "hot colors, hot vice, shady dealings under the palm trees," as Joan Didion characterized Miami in the 1980s.[121] However, throughout the book I explore how, in recent years, the city's image has shifted, in some quarters, toward the apocalyptic, with computerized visualizations of looming coastal inundation now concomitant with promises of waterfront luxury. The new imaginary of Miami as climate change ground zero is a problematization being stitched together via visual and discursive narratives, and it is worthy of analysis in its own right as part of the city's imaginal legacy. But there is also more. For critics, Miami's spectacular nature has long been cause enough to dismiss the city. For Le Corbusier, writing after a 1950 visit, Miami was "enough to make one utterly sick from so much artifice." "Social relevance is not at stake here," the Swiss-French architect concluded.[122] But in the approach I take, Miami's entanglement with images, imagination, and artifice is not something to dismiss but an integral part of the city's material and libidinal reality, and moreover part of what makes the city enthralling, complex, and unique. Miami harnesses this image-reality connection, plays on and elaborates it. And so it is also important to understand how the city's new Anthropocenic imaginaries are bound up with urban planning and thinking at this particular historical moment. Exploring the Anthropocenic imaginaries emerging in and around the city is necessary if one wishes to understand Miami, here, now, at this historical moment—in the third decade of the twenty-first century. Equally, I will argue that pushing beyond the resilience imaginary that dominates Miami and critical thought and practice more broadly depends on reclaiming the basic human capacity for free imagination: a key task for the Anthropocene.

Infrastructure, spatial imaginaries, and imagination are ultimately intertwined. This is one of the more interesting points made by critical infrastructure thinkers: alongside their "background" role, infrastructures also inspire powerful visions and fantasies themselves in diverse relation with liberalism's political imaginaries. As Larkin puts it in an influential essay, poetic as well as technical, modern infrastructures enabled new experiences and perceptions, transforming imaginaries and forwarding powerful desires and dreams.[123] These are entwined with political, economic, and even military dimensions. Throughout this book I explore how infrastructure and spatial imaginaries intertwine in Miami, where they are productive of and producing context-specific problematizations of how the problems of climate change are understood and what solutions are appropriate, necessary, and possible (as well as which ones are not). I explore infrastructures of climate adaptation, but through these I also explore diverse actors' visions of the city's future and spatial imaginaries to which they are tied. In Miami, I will suggest, these two dimensions (infrastructure and imagination) are converging toward a vision and a material space of human and nature as part of a cybernetically interlinked critical urban infrastructural system (albeit one filled with many contradictions). But, equally concerned with solidifying regimes and imaginaries as with their undoing, many of the chapters in this book conclude with a reversal: an exploration of potential unanticipated or currently unthought trajectories opened within specific urban adaptation infrastructures or imaginaries. These possibilities are also the focus of this book, and a key element of what I describe as Anthropocene critical urban theory/practice in chapter 5.

Book Outline

The book is organized as follows. Chapter 1 brings visual and critical discourse analysis of media and policy documents and site observation together with critical work on urban climate experimentation to examine how the reimagining of Miami as climate change ground zero in the United States intersects with the city's political-economic logics. Focused on "Miami Beach Rising Above," a suite of experimental sea-rise and flooding infrastructures—street elevations, seawalls, and pumping systems—under construction in affluent Miami Beach neighborhoods, the chapter examines the affective investments and future imaginaries contained in this large-scale, top-down experiment. Although this project is widely celebrated for its efficacy at fighting urban flooding, this chapter suggests that it also constitutes a highly effective means of counteracting the spectacular media and artistic imaginary of the city as climate change ground zero, and specifically this imaginary's potential impact on urban property values, tourism, investor confidence, and credit ratings. Urban experimentation here is theorized as a form of what the chapter, drawing on resilience ecology, calls "back loop urbanization," a conservative mode of innovation to maintain systems identity amid disturbance. While neither transformative nor progressive, "Miami Beach Rising Above," I argue, is *utopian*. A signal from the climate change front lines addressed to government, finance, and real estate interests in coastal cities worldwide—as well as a recent wave of Silicon Valley transplants working to transform Miami into the new tech and cryptocurrency capital of a post-Covid-19 world—the experiment offers hope that maintaining a certain, historically specific set of socioeconomic urban systems and way of life amid sea-level rise (real and projected) is possible. Conducted, however, in the more-than-human lab of an Anthropocene city, the imaginary embedded in the experiment, the chapter suggests, also struggles with local socio-ecological contexts.

Whereas chapter 1 explores top-down climate experimentation by and for the "most glamourous city in America," chapter 2 analyzes imaginaries operative in attempts to incorporate the city's poor and marginalized into urban climate resilience planning. Drawing on participant observation in and organization of future scenario exercises conducted in Miami, the chapter explores how "seeing like a state" is being transformed in the Anthropocene and theorizes the spatial imaginary and political possibilities of new modalities of Anthropocene state sight. The first, found in governmental efforts to envision futures scenarios, enrolls marginalized urban dwellers to see themselves as government, managers of an entangled, crisis-filled, but still governable city. This is a form of sight that disconnects subjects from their own worlds and capacities and delimits futures to existing governmental images of the present. The second form of sight, found in community disaster scenario games but also Anthropocenic thinking such as that of Bruno Latour, reconnects subjects to their worlds and focuses on building their adaptive capacities for disaster survival. Engaging anthropologist AbdouMaliq Simone's work on "people as infrastructure" and political theorist Julian Reid's critique of governing imaginaries, the chapter argues that while the latter form of sight resonates with contemporary governmental efforts to shape the urban poor into a vital new urban management system—"social infrastructure," a new definition of infrastructure and a vital component of the eco-cybernetic resilient urban apparatus—other trajectories can be found in such scenario visioning efforts. The enframing of human capacities and practices into urban infrastructure, the chapter concludes, both closes and opens possibilities for deep transformation in the Anthropocene.

Whereas the previous chapters analyze infrastructural solutions that were either human-made or made of humans, chapter 3 draws on interviews, site observation, and visual and discourse analysis conducted as part of the Florida Coastal Everglades Long Term Ecological Research Project to examine the reimagining and reengineering of nature as infrastructure. Building on the seminal work of anthropologists Laura Ogden and Jessica Cattelino, this chapter traces the recent revaluation of the vast Everglades wetland ecosystem adjacent to Miami as a valuable urban resilience infrastructure needed to mitigate saltwater intrusion into urban drinking water supply and explores the "geopolitics" enacted across discourse, media, and ecological engineering necessary to make nature function as infrastructure.[124] To distinguish this particular rendering of nature as infrastructure from past views of city/nature relations as well as recent ideas of neoliberal nature and ecosystem services, the chapter forwards the concept of "Anthropocenic infrastructural nature" to describe the contemporary operationalizing of nonhuman flows and processes once thought antithetical to urban development into the city, to administer and secure urban life. This demarcation of threatening, unnatural Anthropocene socionatures like sea rise and what are perceived as presocial, natural natures produces an additional powerful effect: what is forwarded as most natural, and thus unquestioned, is the very urban system itself. But if modern infrastructures are, to use Michael Truscello's word,[125] "necropolitical," Anthropocenic infrastructural nature takes on an additional dystopian dimension in that the natural powers on which urban security now relies may be undermined by the effects of the very industrial, capitalist relations such governance techniques seek to secure.

The technological, human, and ecological infrastructures traced in the preceding chapters are together seen as critical components of a broader urban dispositif—the resilient city—organized around a rarely questioned conviction: even while seas will rise and waves will batter shorelines, the city can be saved. In contrast, chapter 4 analyzes a recurring idea circulating among Miami planners, architects, politicians, and scientists—that the city is already doomed to be swallowed by the sea—and variations on attendant *extreme* adaptation proposals to repurpose and transform it into elevated islands: self-sufficient, artificially designed territories delinked from global energy and production networks. Questioning imaginaries of the Anthropocene as an urban age of endless urban resilience, the chapter argues that these visions of islandizing South Florida, while distinct in approach and ethos, push us to also consider the possibility of an "urbicidal Anthropocene"—understood in a very different sense than this term's typical meaning—as well as the *possibility of islands* in the twenty-first century. Here I conceptualize two forms of Anthropocene urbicide: ending a seen-as-already-doomed urban agglomeration—the end of the resilient city—and breaking from planetary urbanization's infrastructural networks. Engaging critical work on urban ecological security, urban resilience, and planetary urbanization, this chapter analyzes islandization proposals and the challenges they pose to ubiquitous but often-unquestioned visions of the Anthropocene as a resilient urban age. Instead of this now-standard future imaginary, islandization imaginaries, I show, gesture toward an alternative trajectory of dismantling urban environments and infrastructurally delinking, rather than interlinking, futures.

Here, it is important to note that I am not suggesting that an urbicidal Anthropocene "age" would replace urban resilience or planetary urbanization (one totality for another). Rather, my claim is that an urbicidal imaginary is ascendant, at least in Miami, not only in movies and art, where urban destruction is common, but in what some policymakers, planners, artists, and architects envision as pragmatic responses to widespread forecasts of the city's inevitable submergence.

Documenting this new urbicidal vision is important and novel in its own right. While there is a growing and powerful critical literature on urban resilience in Miami—particularly that of Kevin Grove, Savannah Cox, Allain Barnett, and Zac Taylor, among others—little critical work has examined designs for more extreme adaptations seen as necessary after resilience is exhausted. But, here, my exploration of islandization imaginaries is also intended to raise challenges and questions for critical urban thinkers and practitioners working on climate change, urban resilience, and planetary urbanization. However specific and hypothetical, my argument is that islandization proposals raise challenges for urban thinkers focused on the urban-Anthropocene nexus that are worth considering seriously, while understanding so-called urbicide as only *one* of many possible futures.

Chapter 5 concludes the book by reflecting on the need for Anthropocene critical urban theory/practice, able to comprehend urban forms and processes emerging at the intersection of climate change and adaptation, as well as how these mutations may point beyond the theoretical and spatial bounds of the contemporary urban condition itself. This is conceived as a research-and-development agenda which, I further suggest, should also be capable of understanding, questioning, and superseding dominant urban resilience imaginaries. Through their repetition and air of authoritative impartiality, urban resilience imaginaries and framings have become a "common sense," providing legitimacy and momentum, giving meaning to the situation of the Anthropocene, and determining the questions to be addressed and solutions considered appropriate.[126] But what the book ultimately argues is that other imaginaries and problematizations are possible, needed, and already being trialed, at least imaginatively, in a frontline city which push beyond the bounds of urban resilience.

Miami's various experimental sea-rise imaginaries and infrastructures intersect and constitute it as an Anthropocene city, which is neither resilient nor doomed but the sum of multiple, often competing and contradictory, trajectories. While the experimental potential that characterizes the city is currently overwhelmingly geared toward resilience and systems maintenance, this chapter discusses how urban thinkers, designers, planners, and ordinary residents might reappropriate this potential as a means of intervening in the epochal moment rather than being subject to it. Here I meditate on how theoretical and practical insights drawn from social infrastructure and islandization can help inspire and propel urban theory/practice in and for the Anthropocene. Drawing on French philosopher Alexandre Monnin's work on "destauration," Anthropocene urban theory/practice is theorized as a democratic and experimental matter of construction but also dismantling able to shed already-obsolete forms and imaginaries, including those as seemingly inevitable as resilience and even the urban as we know it itself.

Returning to Brenner's challenge to urban thinkers, what I ultimately argue in this chapter and throughout the book itself is that the mapping—and launching—of contemporary transformations in urbanization has not been taken far enough. Contributing to thinking on the broader problematic of what Lauren Rickards, Brendan Gleeson, Mark Boyle, and Cian O'Callaghan call "urban concepts under stress," my aim is to carry Brenner's call for abandoning inherited frameworks and pursuing relentless theoretical and practical experimentation further.[127] Neither this book nor this final chapter is intended as a guide, but rather a call for an experimental disposition, a willingness to abandon inherited imaginaries in favor of exploring others as well as on-the-ground problems and realities encountered by different thinkers and practitioners in their own contexts, as the Anthropocene touches down differently across place and time. As such this approach is about moving away from the liberal "one-world world" of thought and action and opening both up to consider the problems of and potential pathways for urban life in the twenty-first century, without the "banisters" (to use Hannah Arendt's term) of Holocene urban thought and action.[128]

To conclude, a few disclaimers are in order: First, as I see it, the term *Anthropocene*—literally "the human epoch"—is uniquely capable of naming the historical moment we are living in, in which the environmental *and* social impacts of this liberal "one world world," dominated by a single figure of industrial liberal life and its foundational forms of thought and action, are becoming manifest: mass extinctions and rising seas, in the Anthropocene, are the companion to contemporary subjective and epistemological dislocations. When I include the term in concepts I forward, such as Anthropocenic infrastructural nature in chapter 4, I do so to indicate not only that these are infrastructures or imaginaries devised and imagined as responses to on-the-ground perceived problems such as sea rise, but also that their contours have emerged as designers, planners, and media work through what they see as the intellectual and epistemological challenges climate change has provoked (such as disruptions to city/nature binaries). Here the important point is that these are not merely *climate change* responses—mere technical solutions—but *Anthropocenic* responses, with historically specific social and philosophical as well as technical and environmental dimensions.

Second, it is important to note here that this book does not fully achieve its own call for Anthropocene critical urban theory/practice but is instead intended more as an overture and invitation for an ethos of thought and practice to come. It aims to lay out a call for scholars and practitioners both to explore and to create forms of urban thought and practice for the present, a time in which the world is thawing in all domains. While the book offers some initial probes at this via the Miami context, creating Anthropocene critical urban theory/practice is not something that can be done even close to exhaustively by a single author or practitioner—and this is part of my point—but is instead something that would have to be sketched out experimentally by many diverse practitioners and thinkers in their own ways and contexts.

Third, Miami's situation—on a porous limestone foundation, unable to stay dry via seawalls, wedded to spectacular media imaginaries—is unique. Caution should be taken in extrapolating developments there to all coastal cities. Nevertheless, Miami is seen by powerful institutions as a climate change test lab, and developments there will undoubtedly influence other urban imaginaries and practices to some degree. "What happens in Miami," stresses the *New York Times*—an influential voice in stitching together dominant narratives of what appropriate climate change response looks like—"will very likely become a case study for other cities and counties facing climate challenges."[129]

Finally, this book does not by any means intend to exhaustively document or analyze the wide range of climate adaptation designs now being pursed and envisioned in Miami. Instead, it critically explores a selected set of infrastructural imaginaries and, through these, asks broader questions about urban life, imagination, and transformative possibility in the twenty-first century. In the Anthropocene it is clear that new modes of thought are needed. This book is just one contribution to much wider efforts to create forms of thought and strategy from and for the now. It is a work of urban geography and theory—whose critical arguments are made in good faith, with constructive aims and a collegial spirit—but is also focused on real questions circulating among urban inhabitants, including but not limited to those living in Miami, which is a city I love and that is neither just GIS-generated streets on interactive sea-level-rise maps nor architectural renderings (though it is very much these), but is also an extremely unique American city that is both chaotic and corrupt, where the euphoria of sunrise beach raves and flashy wealth of Lamborghinis exists alongside the round-the-clock work and commuting to make ends meet that shapes the working class and poor life in the city, where rather than worshipping the sun from the deck of a multimillion-dollar yacht jetting across Biscayne Bay, one is equally likely to be found stuck in traffic, blood pressure rising and late to work, waiting at the foot of a bridge raised to allow those yachts to pass beneath. Meanwhile, parrots continue to squawk from the imported palms. Enormous thunderheads keep rolling over the city skyline. Seas continue to rise.

1
MIAMI IMAGINARIES
CLIMATE CHANGE GROUND ZERO OR RESILIENT "CAPITAL OF COOL"

> In this mood Miami seemed not a city at all but a tale, a romance of the tropics, a kind of waking dream in which any possibility could and would be accommodated.
> —Joan Didion, *Miami*

Know Your Risk

In the nineteenth and twentieth centuries, much of South Florida was covered in wetlands and mangrove swamps which planners and government viewed as an impediment to urban development, profit, and productivity. To urbanize the region, millionaire investors and the State of Florida conducted large-scale drainage and fill projects to reclaim the "derelict landscape" for commercial agriculture and real estate development.[1] Thus was created the ground on which much of Miami is built. But along with drained wetlands, Miami is equally built on manufactured imaginaries of a subtropical urban paradise. Indeed, it is hard to overemphasize the interconnection between ideas of Miami's habitability and imaginative speculation. Consider, for example, the boosterist images in newspaper ads and tourism brochures of swaying palm trees: a central strategy for attracting real estate investment and development to Miami Beach, which by the early 1920s, as historian Gregory Bush recounts, was for millions of

Americans "associated with speeding cars and motorboats, spectacular stunts, grandiose hotels, mansions, lavish parties, polo matches, bathing casinos, and crowds of sunbathers"[2] (see Figure 3). The spectacle of a decadent, unscrupulous, over-the-top oceanfront city—which historically existed alongside Jim Crow segregation, antisemitism, and extreme economic inequality—remains central to Miami's neon-lit reputation.[3] Today Miami is known as a "a sunny place for shady people," an image based in the city's corrupt reality but often equally derived from popular culture such as *Scarface, 2 Fast 2 Furious,* and *Grand Theft Auto: Vice City.*

But as flooding related to sea-level rise has become more frequent in recent years, speculative imaginaries of Miami have shifted, at least for some, toward a focus on the environmental and technical risk of inhabiting the city and an emerging image of it as *climate change ground zero.* Images of waterfront luxury now compete with a new set of journalistic and social media pictures that capture the city's floods and attempt to raise climate risk alarms. In this vein, a photograph of an octopus found floating in a parking garage during a flood event was circulated widely by media outlets, which described the stranded cephalopod as the city's climate change "canary in the coal mine."[4] Likewise, instead of aerial views of sunbathing vacationers beside the sea, climate change imagery is routinely featured by media outlets like *Business Insider,* the *Guardian,* and the *New Yorker.* Just as commonly, government websites include drone images of the city's canals and waterways, suggestively inviting their viewers to reinterpret these aquatic urban surrounds as the source of impending crisis.[5]

Through the repetition of such images, the city as a hazardscape of multiple, interconnected social, environmental, and technical threats has become a common pictoral trope. Not limited to visual narratives, media and city reports reframe Miami's geology itself as a threat, offering extended explanations of the city's porous limestone foundation (it's like Swiss cheese; the water moves right through it, both horizontally and vertically)

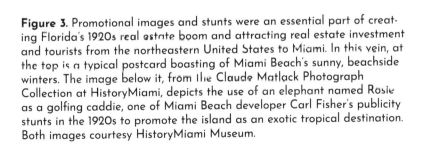

Figure 3. Promotional images and stunts were an essential part of creating Florida's 1920s real estate boom and attracting real estate investment and tourists from the northeastern United States to Miami. In this vein, at the top is a typical postcard boasting of Miami Beach's sunny, beachside winters. The image below it, from the Claude Matlack Photograph Collection at HistoryMiami, depicts the use of an elephant named Rosie as a golfing caddie, one of Miami Beach developer Carl Fisher's publicity stunts in the 1920s to promote the island as an exotic tropical destination. Both images courtesy HistoryMiami Museum.

and the unique sea-level-rise challenges this porous substrate introduces for planners—in particular the fact that seawalls are inadequate to keep out intruding aquatic surrounds. Atop a red-orange vulnerability map of the city's coastline, Miami-Dade County's website implores visitors to discover their neighborhood's risk, linking them to maps created by the National Oceanic and Atmospheric Administration where they can view their area's flooding, storm surge, and sea-rise risks.[6] Even political campaign mailers sent out before elections now include sea-level-rise risk warnings.

While these visual and discursive narratives reproblematize the city's present, others paint a cautionary picture of its coming aquatic future. In the early 2000s, geoscientist Peter Harlem laid out what is now a defining, iconic visual depiction of Anthropocenic Miami: the light detection and ranging (Lidar) sea-level-rise projection map showing South Florida increasingly underwater by mid-2100.[7] Harlem's maps were a critical part of the early formation of Miami's problematization as climate change ground zero, and they generated extensive local and international media attention. In 2013 and 2014, media outlets including the *New York Times*, *Miami Herald*, and *WLRN* published a wave of articles featuring Harlem's maps and apocalyptic titles such as "Millennials Are Flocking to Miami, Where They Might All Drown Soon," as one 2014 *Fast Company* piece put it. In an article titled "Miami, the Great World City, Is Drowning while the Powers That Be Look Away," the *Guardian* dramatically interpreted the implications of Harlem's maps:

> These show that by the time oceans have risen by four feet—a fairly conservative forecast—most of Miami Beach, Key Biscayne, Virginia Key and all the area's other pieces of prime real estate, will be bathtubs. At six feet, Miami city's waterfront and the Florida Keys will have disappeared. The world's busiest cruise ship port, which handles four million passengers, will disappear beneath the waves.[8]

Around the same time, geologist Harold Wanless employed similar sea-level-rise projection maps in presentations warning of South Florida's apocalyptic future (Figure 4). In one such influential presentation delivered in 2008 for the Miami-Dade County Board of County Commissioners, Wanless casts the subtropical city in global, multiscalar connections.[9] It is not merely a global metropolitan area connected by airports, seaports, and highways to international flows of people, capital, and commodities, but also to global earth systems and environmental transformations in distant, previously seen as disconnected, places. The future of urbanized subtropical South Florida, Wanless's presentation suggests, depends on what happens in the snow-and-ice-covered Arctic. Instead of palm tree or beach views, Wanless walks the audience through satellite imagery of lakes, rivers,

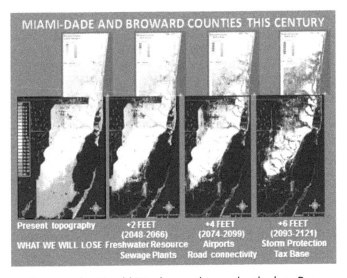

Figure 4. Diagram by Harold Wanless and maps by the late Peter Harlem showing Miami-Dade County's topography in 2014 and the envisioned impacts of two, four, and six feet of sea-level rise on it. Widely distributed by Wanless and Harlem in numerous writings and presentations, such maps and presentations were an integral part of the early problematization of Miami as climate change ground zero in the 2010s. Copyright: Harold R. Wanless and Peter Harlem, used with permission.

Figure 5. Leandro Erlich, *Order of Importance,* 2019. Site-specific public installation of sixty-six life-size sand sculptures of cars and trucks. Commissioned by the City of Miami Beach, curated by Ximena Caminos in collaboration with Brandi Reddick, Culture Affairs Manager at the City of Miami Beach, December 1-15. Photo by Greg Lotus. Courtesy Leandro Erlich Studio.

and moulins opening up and pouring through the Greenland ice sheet and close-up views of dwindling ice cover. As Greenland and Arctic ice melts, he explains, waters around South Florida will rise. This is a process that will accelerate nonlinearly, he emphasizes, as reflective ice surface is destroyed and oceans absorb more solar radiation.

In 2011 the region's sea-level-rise projections were tempered and formalized when the Southeast Florida Regional Climate Change Compact brought together scientists and government to issue the region's first "Unified Sea Level Rise Projection for Southeast Florida."[10] The goal of this collaborative effort was to streamline a single regional sea-rise projection and facilitate adaptation planning in the heavily urbanized area. In the Work Group's updated 2019 forecast, South Florida's future is rendered in a chart depicting lines rising upward: 10–17 inches of sea-level rise (above Key West's 2000 mean sea level) by 2040; 21–54 inches by 2070; and 40–136 inches by 2120.[11]

Drawing on such future-oriented projections in a 2013 *Rolling Stone* feature originally titled "Goodbye, Miami"—which, like Harlem's and Wanless's maps, was pivotal early on in helping to reproblematize the city as climate change ground zero—journalist Jeff Goodell envisioned Miami's twenty-first-century fate.[12] Salt water contaminates drinking water supply. Investors and tourists leave. Roads become undrivable at high tides. Banks stop issuing mortgages. By 2100, "rising sea levels will turn the nation's urban fantasyland into an American Atlantis . . . a popular snorkeling spot where people could swim with sharks and sea turtles and explore the wreckage of a great American city."[13] Drawing on such assessments, much climate-change-themed art produced in or about Miami over the last decade similarly implores audiences to envision the city's twenty-first-century endgame: people playing guitars or drinking coffee while floating in tanks of water; lawn placards indicating homes' future sea-rise risk; a sand-carved traffic jam installed along Ocean Drive (Figure 5).[14] Echoing media and scientific narratives while helping produce Miami's problematization as climate change ground zero, these and many other climate-change-themed artistic productions envision the city and its residents overwhelmed by nature's powerful forces. Concluded Wanless in the widely circulated *Rolling Stone* piece: "Miami, as we know it today, is doomed. It's not a question of if. It's a question of when."[15]

Rising Above

In 2021 Miami-Dade County released its first Sea Level Rise Strategy. From the Tequesta to Gladespeople, and from white and Black settlers to the Miccosukee and Seminole Indians, inhabitants of South Florida, explains the SLR Strategy, have always lived with water.[16] According to the document, the challenge of residing in the region up to 2060 is as follows: "How can we gracefully, strategically live with two feet of additional sea level rise?" To do so the strategy proposes an adaptation toolkit composed of five approaches devised for the region's unique communities and geological conditions: "Build on Fill," "Build Like the Keys," "Build on High Ground Around Transit," "Expand Greenways and Blueways," and "Create Green and Blue Neighborhoods."[17]

Much in Miami-Dade's living-with-water toolkit is prospective. However, some of these strategies are already being trialed in Miami Beach, a low-lying, seven-square-mile barrier island flanked by the Atlantic Ocean and Biscayne Bay, where the City of Miami Beach for some time has devoted itself to the task of building climate resilience. Now labeled a "living laboratory" for urban resilience, Miami Beach has taken up the resilience mantle in diverse, highly publicized ways, including joining the (now defunct) Rockefeller Foundation 100 Resilient Cities and the Greater Miami and the Beaches collaboration, participating in Columbia University's Center for Resilient Cities and Landscapes' 100 Resilient Cities Resilience Accelerator program, and hosting the Aspen Ideas: Climate Conference, a multiday even that in 2023 brought business leaders, activists, and politicians such as Vice President Kamala Harris to the city to discuss climate adaptation solutions and tour the city's resiliency projects.[18]

Alongside and entwined with its marketing efforts, Miami Beach is currently the site of a projected $650 million, ten-years-plus experimental climate resiliency infrastructure program laid out under former mayor Philip Levine, a "cruise line media magnate" elected in 2013 on a campaign promising to build resiliency to current and future floods, even filming one of his campaign videos as a gondolier, leaving work by paddling through flooded streets in the back of a truck.[19] Under Levine, via no-bid emergency contracts, doubled stormwater fees, and multiplied sea-rise projections, the City of Miami Beach fast-tracked a suite of stormwater management infrastructure projects, including installing a fleet of industrial pumps, elevating streets, and building new seawalls designed to prepare Miami Beach for sea-level rise, and specifically coastal flooding, King Tides, heavy rain- and storm-related flooding, and groundwater flooding.[20]

As part of the plan, referred to as "Miami Beach Rising Above" and continued in an expanded form under Dan Gelber, Miami Beach's mayor from 2017 to 2023, the city is raising streets throughout the island to mitigate sea-level rise, rain, and high-tide flooding projected to the year 2060.[21] The original plan called for raising roads to stand at a minimum of 3.7 feet above NAVD 88 (North American Vertical Datum), and a revised strategy, created after resident protests and expert consultations, takes a more flexible approach, recommending varying elevations for emergency, major, and residential roads.[22] In the road-elevation project's initial pilot neighborhood, Sunset Harbour, engineers elevated all required roadways up to three feet.[23] The neighborhood is located on the island's western bayside and is home to a cluster of expensive condos and restaurants. In the latter, seating is now located two feet (two feet of sea-level rise is the upper limit of the county's SLR adaptation strategy and is projected for approximately 2060) below the recently raised street. Along the elevated roads, one finds upscale Mediterranean bistros, Botox providers, and fitness boutiques. With Sunset Harbour's elevation considered a success, road elevations have been expanded across the city to areas including the exclusive Palm and Hibiscus Islands as well as Bay Road, West Avenue, and North Beach.

Existing businesses have been modified to accommodate the elevated streets. For example, the Sunset Harbour Publix, a grocery store chain ubiquitous throughout Florida, was built to what several years ago was base flood elevation, with seven stairs leading from the previous street height into the store's entrance, has now had five of the previously existing stairs leading into the large store eliminated, with just two stairs now leading into the store.[24] Complementing the raised roads and structures, Miami Beach has torn up streets to install more than thirty $2–3 million industrial pumps with plans to install dozens more in coming years to funnel floodwater out of city streets and into adjacent Biscayne Bay.[25] Capping off the "Rising Above" plan, the city is installing a series of new seawalls to protect streets and property from flooding coming from the bay. The aim is that by eventually raising the entire city, Miami Beach will gracefully maintain function and identity as seas rise.

Urban Experimentation

Miami Beach's resiliency efforts are seen as an experiment, with the city celebrated by government and planners as a living laboratory for testing climate change solutions. Here the city is part of the broader trend in which urban laboratories have become an increasingly privileged practice for urban adaptive governance in pursuit of resilience.[26] Urban governance has long been experimental.[27] But as climate change has emerged as a key security concern and as urban government shifts from mitigation to management amid unmitigated change, planners, designers, and municipal bodies have begun arguing that government in the age of climate change requires experimentation with new methods in real-world, live urban contexts.[28] This governmental embrace of urban experimentation—or "governing by experiment," as geographers Harriet Bulkeley and Vanesa Castán Broto put it—is frequently tied to the recent discrediting, often inspired by the grappling with the Anthropocene's impacts, of modern governance's ideas of control and certainty.[29] For many, the latter are now seen a failure, the very cause of climate change, environmental degradation, and network fragility.[30] As geographer Simon Dalby argues, modern governance techniques have themselves generated new risks—such as climate change—which threaten the very liberal life those modes of government sought to secure.[31] To protect not only biological species life, as in past biopolitics, but also the biophysical and infrastructural conditions of that life, urban governments now search for experimental new ways to secure the environmental, infrastructural, and social lifelines of liberal society.[32]

Often construed as breaking from modern control-based governance, urban experiments are portrayed as innovative ways to generate new knowledge about the city, to try "provisional, risky, and dynamic" governance experiments that embrace uncertainty and learn by doing, and "are used to explore uncharted policy territories, to either learn or open up new forms of intervention, and often without consideration of their unintended consequences."[33] A growing number of critical scholars are now exploring urban climate experiments, addressing projects such as Manchester, UK's Oxford Road corridor low-carbon experiment; Baltimore's Long-Term Ecological Research project; and low-carbon housing developments in Bengaluru, India, and Monterrey, Mexico, to name a few.[34] Highlighting the ubiquity of urban experiments across different world regions, Broto and Bulkeley have identified 627 climate change experiments in 100 global cities, including carbon sequestration, urban tree planting, and landfill gas capture.[35]

Miami Beach's infrastructures are part of this shift toward governing climate change via experimentation, albeit in ways that differ from many other urban climate trials. As characterized by James Evans, for example, urban climate experiments are the result of a shift from modernist control-based approaches and toward adaptive management of cities as complex, integrated socio-ecological systems.[36] Here urban resilience emphasizes the need for designs based on socio-ecological systems thinking and entangled human-environmental flows.[37] In this vein we might think of New York's use of oysters as "living infrastructure," coproduction-based projects involving multiple stakeholders, or other nature-based solutions and green infrastructure efforts often led by nongovernmental actors.[38] These recalibrated systems-based resilience approaches will be discussed in chapters 2 and 3. In contrast, however, to these iterations of the experimental orientation, Miami Beach's venture is a standard top-down, large scale city-led gray infrastructure project using cement and large machines defined in its early days by local newspaper *Miami Herald* as an "engineering-first, green-solutions-never approach."[39] As will be discussed further in chapter 3, such resident, media, and external consultant criticisms have led the project to incorporate more nature-based design as it moves forward; nevertheless, by and large the focus remains on the gray infrastructure components.

Instead of rethinking modern typologies, then, the experimental nature of "Miami Beach Rising Above" has more to do with factors such as the speed and scope of implementation as well as the fact that the program is understood as testing new solutions within unprecedented conditions of rapidly encroaching sea rise in a heavily developed urban area. Described as part of a necessary race against time, the infrastructures, at least initially, were laid out with "unusual speed"[40] over a multiyear flurry, during which Levine staged several spectacular press events promoting the designs' success in keeping streets dry in Miami Beach during King Tides, as well as relentlessly in magazines such as *Rolling Stone* and as the face of climate change first response in Leonardo DiCaprio's climate change documentary *Before the Flood*.[41]

Moreover, urban climate experiments are often described as small-scale, site-specific.[42] But rather than small-scale demonstration sites—what Andrew Karvonen, James Evans, and Bas van Heur call "niches," or what Raymond De Young and Stephen Kaplan call "small experiments"—Miami Beach's infrastructures work at a larger scale to transform urban infrastructure.[43] The city is elevating over 100 miles of streets with the eventual goal being to raise the entire city gradually over time—an audacious scope and approach that has not been tried before.[44]

Taken together this way in which the city is understood as testing new, untried solutions at a grand scale within unprecedented conditions is the most fundamental way in which its resiliency program is viewed as an experiment. As Levine put it,

> When you become mayor of a city that's being attacked by sea level rise, there's no book to buy, no course to take, no guide to follow. We had to literally write the book ourselves. . . . We have learned as we went along. We of course realize that

everything we did could have been better and everything we've done can certainly be improved.[45]

Is it viable to elevate an entire modern city? Will pumps function as expected? Echoing Levine, Elizabeth Wheaton, the city's former director of environment and sustainability stated, "We don't have a playbook for this."[46] Instead, projects are evaluated as they are built, in tandem with shifting projections of sea-level rise and climate change, as well as evolving ways of defining appropriate responses. In the process the city is seen as writing a new climate change playbook in real time, one that will be a model for other coastal cities moving forward.

In the many think pieces and articles on Miami Beach's situation, such designs are highly regarded for all these reasons. A 2018 panel convened by real estate and land-use think tank the Urban Land Institute (ULI) commended the city's resilience efforts and has praised it for having the "courage" to act on them.[47] ULI further advised the city to continue "embracing the resilience brand" in order to distinguish itself from other, competitor coastal cities, communicate itself as a proactive resilience leader locally and internationally to residents and business, and even explore developing the concept of "living with water" as part of the city's "tourism and economic development branding as well as its resilience and stormwater management strategies."[48] Former special envoy for International Water Affairs of the Netherlands and influential Rebuild by Design principal Henk Ovink called the infrastructures "exemplary" and "the way to do it."[49] In his *Years of Living Dangerously* special on Miami, even comedian Jack Black "applauds [the mayor's] efforts."[50] In Levine's own retrospective appraisal, one repeated often, the projects may be flawed but are necessary to prevent the city from becoming the next Atlantis.[51] Instead of a doomed future fossil, Miami Beach is #RisingAbove, as the city's marketing campaign puts it in a slogan that adorns the city's sun-faded bus stop coverings, with the increasingly elevated roads referred to as the "streets of tomorrow."[52]

As Bulkeley and coauthors emphasize, however, more is often at stake in urban living laboratories than mere technical experimentation. Indeed, they write, "experimentation provides a means through which diverse actors seek to navigate and make sense of the present whilst also giving concrete form to particular visions of the future."[53] With this in mind, one might ask, What way of navigating and making sense of the present drives this experimental effort to rise above? To what perceived problems do these pumps, and seawalls, and elevated streets respond? And what visions of the future are driving and being concretized in them?

Economies of Resilience

In cities like New York, resilience projects are often presented as a dynamic and revolutionary new design approach based in complex, entangled socio-technical-ecological systems.[54] As we will see in chapters 2 and 3, these ideas are also present in other resiliency infrastructures being envisioned for Miami. However, in Miami resilience is at the same time often understood in a directly economic register, oriented toward maintaining real estate markets, tourism, investment, and credit ratings in Southeast Florida. For Levine, Gelber, and many others, a key goal of resilience infrastructure building in Miami Beach as well as across Miami more broadly is to maintain urban real estate markets.[55] This entwinement between climate adaptation and property values is clear in the city; as the *Miami Herald* put it in a 2020 article about Miami Beach's road elevations, "At stake is the city's ability to survive climate change, both the two feet of sea rise expected by 2060 and the potential loss of property value in risky coastal areas like Miami Beach."[56]

This concern is only logical if one considers the city's past and present real estate economy and the way it has long been and continues to be tied to speculative imaginaries of the city— whether those of subtropical urban paradise, climate change ground zero, or the new post-pandemic imaginary being sketched out alongside both, which I will discuss momentarily. After all, Miami was founded on long-distance real estate speculation—itself propelled by boosterist ads depicting the city as subtropical paradise—and throughout various boom and bust cycles, Miami has grown into a global city operating as a growth machine post–Great Recession, driven in particular by Latin American investment.[57] As capital transfers out of politically unstable regions of South America, assets from Russia, and dirty money of all varieties find investment homes in Miami's real estate markets, they have contributed to ever-skyrocketing housing prices and extensive new development throughout the metropolitan area, despite growing public discussion of sea-level rise.[58]

These trends exploded during and after the 2020 Covid-19 pandemic, which according to sociologist David Madden has "inaugurated a new phase of urban political-economic recomposition, realignment, and restructuring."[59] Driven by a range of political projects initially legitimated in the name of pandemic response, this emergent paradigm of "covid-capitalist urbanization," as he names it, is characterized by the escalation of preexisting forms of rentierism, the remodulation of workplace conditions, and the intensification of the relationship between urban government, tech companies, and algorithms. One might think here of how the pandemic has served as an opportunity for live trials of surveillance and tracking technologies in the form of selfies, robotics, and AI, as well as "pandemic smart city" tech such as contact tracing apps and drones, or social contact monitoring and vaccination pass systems.[60]

However, the pandemic also became an enabling environment in which other urban governmental experiments, particularly with manufacturing new urban imaginaries around tech, finance, and cryptocurrency, have been launched. Such trials and their impacts on real estate have been especially prominent, and dramatic, in Miami, which along with being reproblematized as climate change ground zero is also, post-pandemic, being reinvented as the new "Wall Street South."[61] In January 2021, in a story reposted by many local real estate and media outlets, the city was ranked the "ritziest, glitziest, and trendiest" city in America:

> Sunny Miami dominates our ranking of the Most Glamorous Cities with more cocktail bars, lounges, and nightclubs per 100,000 residents than anywhere else in the U.S. But it's in the beauty, fashion, and shopping category that Miami really shines. Miami took the top spot in a whopping four shopping-related metrics: luxury fashion, jewelry, watch, and perfume shops. There's no better place in American to party hard and spend on bling than The Magic City.[62]

Similarly, in its 2022 "World's Greatest Places" issue, *Time* declared Miami "so hot right now" and "the U.S.'s capital of cool."[63] The *Financial Times* named Miami "the most important city in America."[64] Amid a Covid-fueled exodus of wealthy individuals and companies from colder, more crowded cities like New York to work remotely from second or new homes in the suburbs and rural areas, Miami was one of the top destinations for Americans relocating during the pandemic. Housing prices have skyrocketed.[65] Fueled by a social media campaign orchestrated by the City of Miami's mayor, Francis X. Suarez—who proclaimed himself the "tech bro mayor"—since 2020 a growing number of tech companies announced their relocation from Silicon Valley to Miami. Suarez's self-described "How can I help?" movement to make Miami the tech, finance, and crypto capital of the country revolves around social media, financial incentives, and a cryptocurrency branding campaign (Suarez widely promoted MiamiCoin, the first "CityCoin" powered by the Stacks protocol—and proposed using it to pay municipal employees), alongside the allure of no state income tax and remote work calls set against real swaying palm trees—not fake Zoom background versions.[66] In a stunt reminiscent of Miami's early days

Figure 6. Tweet by Miami mayor Francis Suarez, unveiling the city's new Miami Bull statue at the opening of the 2022 Bitcoin conference.

of publicity gimmicks—elephant rides, race cars, billboards proclaiming "It's June in Miami" on display in New York's Times Square, and so forth—at the opening of the 2022 Bitcoin conference held at the Miami Beach Convention Center, Suarez unveiled the city's new mascot: a massive, "ultramodern robotic" bull statue with Bitcoin laser eyes. Declared Suarez of the play on Wall Street's famous bull statue, "The future of finance is in Miami and the Capital of Capital now has its own bull 🚀"[67] (Figure 6).

MiamiCoin lost 90 percent of its value within a year of launching and was suspended from OKCoin exchange just months after the FTX collapse.[68] However, other varieties of cryptocurrencies, including Bitcoin, remain healthy, and the post-Covid rebranding of Miami has exacerbated the city's already extremely unequal real estate market, producing skyrocketing home prices with tech investors like Keith Rabois (a vocal promoter of the tech relocation who calls sea-level rise "fake news") paying $29 million for a 15,000-square-foot Miami Beach waterfront mansion, outfitted with a massive 5,600-gallon saltwater aquarium, the maintenance of which requires a scuba diver.[69] PayPal cofounder Peter Thiel purchased two neighboring waterfront mansions for $18 million, and in 2023 Amazon founder and (at the time) third-richest person in the world Jeff Bezos purchased two neighboring mansions for $79 and $68 million on Indian Creek—an exclusive island also known as Billionaire's Bunker— and announced his permanent relocation to Miami.[70] In 2022, billionaire Ken Griffin's hedge fund Citadel and market-making firm Citadel Securities moved their global headquarters to Miami.[71] Investment and financial management companies such as Blackstone, SoftBank, CI Financial, Thomas Bravo, and Novo, as well as the crypto company Blockchain.com, also moved their headquarters or offices to the Miami metro area. To service the newly arrived population, New York restaurateurs are opening Miami locations, and waitlists at private schools and country clubs are growing rapidly. According to RealtyHop, in 2022 Miami became the most expensive housing market in America.[72]

All of this is alongside and intertwined with Miami's longstanding tourist economy, on which the city is heavily dependent and whose post-pandemic growth was propelled at least in part by these rebranding efforts. Florida overall was the destination of 137.4 million visitors in 2022.[73] In Southeast Florida a great deal of tourism is focused on Miami and Miami Beach, and despite pandemic disruptions, in 2021 Greater Miami and Miami Beach hosted 24.2 million visitors, who accounted for $19.2 billion in visitor spending.[74] Fifty-three percent of these visitors stayed in Miami Beach.[75] These numbers increased again in 2022, when 26.5 million came to Greater Miami and Miami Beach and spent $20.8 billion, 8 percent more than in 2021.[76] Total sales and tourist-related taxes collected for Miami-Dade County in fiscal year 2020–21 were over $1.5 billion.[77]

Resilience as Crisis Response

In Miami, two urban imaginaries—"climate change ground zero" and the United States' "capital of cool"—thus coexist. Both are simultaneously real—experienceable on the ground—as well as speculative. And, returning now to Miami Beach's resiliency efforts, we can see how the success of the latter imaginary (Wall Street South, the capital of cool) depends on managing and mitigating the impacts of the former (climate change ground zero, doomed next Atlantis).

Indeed, the economic threats posed by sea-level rise—both current and projected—to Florida and Miami's real estate markets, as well as its tourist industry, investor and insurer confidence, and credit ratings have been at the center of media, academic, business, and planning discussions regarding climate change and resilience in recent years. In a much-cited Risky Business Project estimate, sea-level rise is projected to plunge $15–23 billion worth of Florida's property under the ocean by 2050.[78] In the report *Getting Ahead of the Curve for a Resilient Economy: An Introduction to Risks and Opportunities of Climate Change for Top Business Sectors in Southeast Florida*, rising seas are forecast to affect all aspects of Greater Miami's key tourist sector, including tourist arrival and spending, as well as tourism-related jobs including hotels, restaurants and bar hospitality, cruise ships, and beach recreation as well as conventions, with the Miami Beach tourist industry alone valued at more than $11 billion spent by 15.5 million visitors to the island each year.[79] According to *The Business Case for Resilience in Southeast Florida*, a report produced by the Urban Land Institute and commissioned by the Southeast Florida Regional Climate Change Compact and associated business communities, by 2040 the Southeast Florida region is projected to see seventeen inches of sea-level rise, daily tidal flooding could threaten $4.2 billion in property value and $28 million in sales, property, and tourism taxes due to permanent sea-level rise.[80] These numbers, they predict, would rise in 2070, with forty inches of sea-level rise, to $53.6 billion lost in property values due to daily tidal inundation, and $384 million in fiscal losses due to permanent sea-level rise.[81] Miami-Dade County specifically, they forecast, will face impacts up to $106.5 billion in property, $361 million in tourism and sales, and $2.388 billion in property taxes over the next fifty years due to sea-level rise and flooding—if no resilience-building action is taken.[82]

These economic projections are constitutive parts of the city's climate change ground zero imaginary, which they help stitch together and solidify. But they are also calls for urban growth coalitions to manage the impacts *of* that imaginary, and to ward off the future it envisions. In many ways, Miami-Dade and Miami Beach's turn to resilience can be understood as a pragmatic response not only to the actual impact of climate change (actual floods and rising seas) but also to the climate change ground zero *imaginary* of the city's future that has been sketched out in the last decade (in "Goodbye Miami"–style media pieces, scientific projections, academic and business reports, and art)—and its potential impact on the city's economic reputation and thus real estate values, tourism, credit ratings, and investor and insurer confidence. As the Urban Land Institute warned in their 2020 study,

> The Southeast Florida region is already facing reputational risk; coastal hazards and the region's vulnerability to sea-level rise have received significant national coverage in the U.S. popular media. Vulnerability to coastal hazards now and in the future can result in reputational risks and associated impacts such as property devaluation, insurance premium increases, bond-rating downgrades and increased borrowing costs, decreased tourism and associated spending, decreased public support, and risk from increased liability.[83]

Thus, while once a subject seen as taboo by South Florida's business community, the economic threats posed by rising seas and need for resilience building have become an increasingly common point of discussion, with prominent Miami real estate mogul Jorge Pérez, for example—whose Pérez Art Museum Miami (PAMM) is elevated twenty feet, made of thick concrete walls and reinforced glass, and features a floodable parking garage—proclaiming, "sea level rise is something that is going to hit us all."[84]

In this context, resilience is increasingly understood as a requirement for continued confidence in Miami's economic health and an important signal to banks, investors, developers, and credit rating agencies that the region remains financially viable. Referring to Miami Beach's road-raising efforts, former mayor Gelber cautioned, "If we don't go forward with this project now, I am convinced there will be a day when an insurer will say I won't approve this property, a bank will say I won't do a loan."[85] Suarez has likewise stated that "if insurers decided to pull out of the city of Miami, our city, our real estate, would grind to a screeching halt. We want to make sure we prevent that from happening, and continue to have responsible development going forward, and protecting our most critical assets."[86] The postpandemic crypto, finance, and tech rebranding of Miami is similarly described frequently as dependent on climate resilience action. A 2023 *Bloomberg* article on Miami's efforts to rebrand itself as the Wall Street South, for example, stipulated three key changes municipal authorities must make if the city is going to "fulfill its promise as a financial center."[87] Alongside investing in schools and walkability, the authors emphasized the need for climate resilience infrastructure, noting that "South Florida's local governments need to aggressively push ahead with climate adaptation plans" and that "failing to do so could deter financial institutions from building out their offices and luring staff to South Florida."[88]

Resilience is now similarly highlighted by credit rating agencies such as Moody's, which in recent years have begun including climate vulnerability and resilience into their rating of municipal bonds. In her work on finance and resilience in Miami, urban geographer Savannah Cox explains that, to do so, analysts consider climate risk projections for a given city and compare these with resilience initiatives, specifically infrastructural ones, on the ground. As one analyst interviewed by Cox explains, "absence of any initiatives or infrastructural improvements makes you more vulnerable to what the science is telling us. If we believe we see a mismatch between what an entity does and the risks they face related to climate change over time, their rating will likely change."[89] Notably, analysts also, Cox shows, consider local media stories to understand local climate- and resilience-related issues—how they are perceived locally and how governments are responding to them.

Seen in this light, that the outpouring of apocalyptic media pieces reproblematizing Miami as climate change ground zero around 2013 and 2014, highlighted at the beginning of this chapter, was a clear concern for the city's governments and a motivator driving the launching of "Miami Beach Rising Above" is not at all astonishing. Negative media pieces are seen by rating agencies such as Moody's as adding to a city's risk profile, whereas positive pieces profiling municipal resilience-building efforts—such as the many documentaries, articles, and reports featuring Levine and the "streets of tomorrow" that accompanied #RisingAbove's launch—are seen by analysts as additional proof of resilience. In this context, to counter media and climate projection imaginaries, governments must produce their own counternarratives to signal to rating agencies (as well as insurers and developers) that they are not, in fact, "doomed" but are instead *resilient*. Likewise, "to get ahead of the curve for a resilient economy," a report from Florida International University's Sea Level Solutions Center created for Miami's business community strategizes ways to create positive public, media, and business perception of resilience projects in order to maintain confidence in the city's business and tourist industries.[90] Forefronting the importance of urban imaginaries of and attitude toward climate change in Miami, scholar of consumer decision analysis and marketing Robert Meyer suggests that, to maintain property values and thus the real estate taxes that help fund the city's resiliency measures, "rather than sounding alarms . . . there is an implicit sense that the best approach may be, ironically, to do the opposite. And while a strong case can be made that this behavior has no rational basis, it may represent Miami's best long-term hope for dealing with the threats posed by climate change, one that other cities might be advised to mimic: the best strategy may, in fact, be to foster a collective belief that there is no threat—or at least not one serious enough to lose sleep over."[91]

Considered from this perspective, what is at stake in resilience efforts such as Miami Beach's elevated roads, then, is not simply a technical response to urban flooding. Rather, underway is also a battle between two competing urban *imaginaries*. To combat the potential impacts of the climate change ground zero imaginary, Miami Beach decision makers simply forwarded a *different* imaginary, that of a resilient, thriving city in which "sunny day flooding [is] a thing of the past!," where educators, planners, and governments from around the world can visit to learn from what is seen as the first successful climate adaptation project in Florida,[92] and where raising streets has not only maintained but also *raised* real estate values while helping maintain credit ratings.[93] In fact, regarding the latter, Miami politicians repeatedly link resilience building to credit ratings, both their maintenance and potential upgrade, with both understood as proof that resilience works. For example, a 2019 report sent from Miami Beach's city manager to Mayor Gelber, cited by Cox, announcing Moody's continued assignment of an "Aa2" (high grade) rating on the city's general obligation bonds emphasizes this:

> I am pleased that we have been able to maintain our strong credit ratings through our proactive efforts to reduce risk by investing in our aging infrastructure and adapting to climate change by using some of the best available science and knowledge. We must continue to act along these lines as climate resilience will continue to be a consideration for future ratings.

It is essential that we protect our tax base and our financial standing by continuing to adapt and remain committed to our resilience policies, programs, and operations.[94]

The report further highlights the role that "Rising Above" infrastructural improvements played in Moody's rating decision:

> Please note that in the attached report, Moody's mentions that "management has invested substantially in raising sidewalks and streets and stormwater and water and sewer infrastructure.... Management includes sea level rise assumptions in all capital planning and will continue to invest in climate change mitigation."[95]

The result, as Cox suggests, is an almost circular dynamic in which (as many Miami Beach residents themselves point out) the city ends up building resilience infrastructure not simply because it works (and it doesn't always) but to improve credit ratings and more broadly its status as investable. A whole new urban infrastructural layer begins to appear, then, nominally for climate resilience and climate adaptation, but also, critically, to maintain the city's economic reputation and demonstrate financial viability. And climate resilience itself—as a goal, a framework, and a practice—once it is targeted and required by credit rating agencies, is increasingly legitimated and dispersed as the best and only response to sea rise and flooding, current or projected. As such, Cox concludes, the definitions of resilience and actual infrastructures built to achieve it are an *effect* of this dynamic, and more specifically "the observational practices of rating agencies."[96]

Much the same might be said of resilience as a response to warnings of a looming sea-level rise and flood-propelled insurance crisis in the region. There is growing concern of an insurance crisis in Florida—amplified in media but also experienced on the ground by residents—which a 2023 First Street Foundation report labeled a coming "climate insurance bubble."[97] In

2023 Farmers Insurance pulled its company-branded policies out of the state, joining Bankers Insurance and Lexington Insurance, which had pulled out of the state the previous year.[98] Meanwhile, AAA announced its decision not to renew some policies in Florida. Many Florida residents are now facing soaring premiums—Miami is listed as having the highest premiums among U.S. cities—where insurance companies have pulled out they are being forced to turn in large numbers to the state's insurer of last resort, Citizens.[99] Farmers Insurance explained their pull-outs or rising premiums on a host of factors, including recent devastating hurricanes, flooding, the vulnerability of dense coastal development, lawsuits, and construction costs.[100]

Indeed, former mayor Levine described the context in which he initially formulated Miami Beach's "Rising Above" infrastructure projects in 2013 as a moment of impending crisis, in which action was seen as critical to avoid soaring insurance premiums, plummeting real estate values, and a possible mortgage crisis. As he put it, "The insurance companies of the world don't really care where the water is coming in. If they don't see that we're making significant efforts to make our city resilient, they'll stop insuring. Finance companies . . . will stop financing 30-year loans. The biggest crisis we can have is a loss of confidence."[101] Levine's resolve was clear. As he explained to staff in an email exchange regarding stormwater management projects on Indian Creek Drive, "we need to get this done on emergency 'turbo' speed," concluding, "EMERGENCY FAST???? Unbelievable sense of urgency?"[102]

Back Loop Urbanization

"Miami Beach Rising Above" represents one among several possible infrastructural and imaginal responses to upheavals of the Anthropocene, a type of response that can be conceptualized through a heuristic that has been central to resilience thinking itself: the adaptive cycle. According to resilience ecologists, all systems go through a cycle with two phases, a "front loop" and a "back loop" (Figure 7). The front loop is seen as progressing from an initial "growth" or "exploitation" phase to a second phase of seeming stability when everything is "in its place." Meanwhile, a back loop is a time when those structures come apart, a period of destabilization, confusion, and fracture, but also potential for experimentation, reorganization, and transformation.[103] Stepping back momentarily from Miami, this heuristic is a helpful lens through which to see our current place in the Anthropocene: as industrial liberal society has pushed Earth beyond several planetary thresholds, we are leaving behind the so-called stability of the Anthropocene's front loop and entering its back loop. The boundary crossings and dislocations of the Anthropocene back loop concern, on one hand, Earth's moving out of the safe operating space of stable climates and abundant, unpolluted resources and biodiversity that were the basis of the Anthropocene front loop's short-lived but once seemingly secure, linear world. While such matters are the standard purview of Anthropocene studies, the back loop, on the other hand, also concerns the human realms of thought and action, including transforming modalities of urban governance, planning, and design devised in *response* to environmental transformations.

As resilience ecologists see it, endogenous and exogenous disruptive events such as fires, floods, or droughts test systems in the stability phase, leading to a back loop time of uncertainty in which institutions and individuals must respond.[104] There are many possible responses to a back loop. Management may take the form of avoidance, or of efforts to stay in the front loop and maintain stable parameters and approaches. In contrast to modes of governance that attempt to stop or prevent risks in advance, resilience practitioners recommend instead a period of experimentation, which embraces uncertainty and uses it as an opportunity to test out new management responses.[105] Such an approach begins from acknowledgment that there is a buffer zone within which regimes can exist while still maintaining their state or identity, and experiments seek to find those thresholds while testing out new ways to maintain systems. In its current resiliency iteration, *back loop urbanization*, as I propose

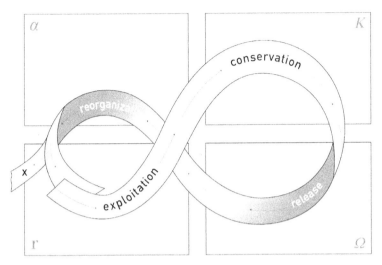

Figure 7. The adaptive cycle, showing a system's two phases of life, the front loop (the light band in front, passing from exploitation to conservation) and the back loop (the darker band in the background, passing from release to reorganization). Adapted from Lance H. Gunderson and C. S. Holling, eds., *Panarchy: Understanding Transformations in Systems of Humans and Nature* (Washington, D.C.: Island Press, 2002), 34, by Caroline Castro, 2019. Courtesy of Caroline Castro.

we name it, represents the taking up of this approach to urban systems faced with already-occurring dislocations such as rising seas and flooding, as well as, more immediately and powerfully, at least in Miami, the potentially catastrophic disruptions caused by media and scientific imaginaries and projections *of* these. Rather than deny the reality of these events or attempt to use past management forms to administer them, back loop urbanization represents the range of experiments underway to remodulate cities and urban spaces so as to govern such transformations as they play out now.

Far from natural or transhistorical, these are forms of urbanization and urban government that are specific to the Anthropocene. Too often, Anthropocene-inspired scholarship uncritically imports metaphors and concepts from the natural sciences into social science work, in the process naturalizing social, political, or technical phenomena. It is therefore important to emphasize that this usage of the back loop concept should not be interpreted as naturalizing urbanization—after all, one of the aims of this book is to position urbanization itself (as we know it) as a historically specific, potentially finite set of (always transforming) processes. With this in mind, the back loop is intended to function as a potentially helpful way of understanding contemporary reconfigurations within urbanization and urban government. As mobilized within resilience ecology, the back loop is a heuristic, not a natural phenomenon, invented by ecologists as a lens through which to interpret and consider complex realities.

As one among multiple possible modalities of back loop urbanization, Miami Beach's resiliency efforts appear as a relatively straightforward, conventional approach to standard resilience, one that is experimental yet relatively moderate in its aims and future vision. Its aim is to respond to current sunny-day flooding while also anticipating near-future crises, in order to preserve existing social and economic systems. While relatively conventional or conservative in this sense, Miami Beach's style of back loop urbanization does put a twist on contemporary forms of anticipatory governance.[106] As Claudia Aradau and Rens Van Munster argue, the increasing omnipresence of catastrophic risks caused by industrial society such as global warming or terrorism has led not to a world of ungovernable and uninsurable danger but instead to new ways of governing future crises amid uncertainty.[107] This dispositif of anticipatory action reproblematizes and transforms past modes of governance, supplementing modern techniques with heterogeneous practices of identifying, calculating, and governing futures.[108] Anticipatory governance practices include precautionary techniques to preempt, manage, or otherwise secure liberal societies against future catastrophes such as terrorism using both exceptional and everyday measures, techniques of "enacting" threats such as nuclear attacks.[109] And with the "climatization of security," mitigation, preparedness, and disaster management have become key practices for anticipating the effects of unmitigated climate change on defense, migration, and development.[110]

In "Miami Beach Rising Above," however, the aim is not only to anticipate and govern actual future threats—sea-level rise, flooding—but *also* to foresee and govern the impact of anticipatory *imaginaries* of ruination (to use Paprocki's term) on current social and economic systems—rather than biological species life or its biophysical conditions—by countering one imaginary of climate risk with another vision of climate *resilience*.[111]

As technologies of anticipatory risk governance, Miami Beach's elevated roads and pumps also have their own relationship to performance and demonstration, one that is perhaps unique within urban climate experimentation. In contemporary risk governance, it is common to perform hypothetical future catastrophes in order to anticipate, learn from, and govern them (such scenario exercises are discussed in chapter 2).[112] However, many climate experiments, Gareth Edwards and Harriet Bulkeley note, frequently take place in the urban milieu and involve concrete demonstrations of other possible futures in the present.[113] In Miami Beach, however, experimental resilience infrastructures respond to present-day events: rising seas and sunny-day flooding, but also, critically, they respond to the anticipatory *performance* and *demonstration* of the city's climate-change-induced future *ruination* within media and scientific reports—and the potential impact of the imaginaries generated by these performances on the city's economy.[114] This is a present-oriented, short-term infrastructural experiment whose function is to administer flooding as well as mediatized climate change doom imaginaries—through the construction of resilient infrastructure and urban imaginaries both.

As one style of back loop urbanization, "Miami Beach Rising Above" thus imagines and seeks to create specific futures, albeit in ways that differ from many other urban climate experiments. The latter frequently imagine futures that are radically different but also qualitatively better than the present—at least in the eyes of practitioners launching the projects—with future climate change projections seen as impetus for imagining utopian tomorrows. As analyzed by Edwards and Bulkeley, for example, Philadelphia's Coolest Block Contest with its energy-efficient retrofitting and Berlin's Solar Atlas rooftop solar energy project both seek transformation toward low-carbon, sustainable, socioeconomically just futures.[115] Edwards and Bulkeley thus argue that climate change experiments are "heterotopias," "spaces of utopian hope" that combine risk with opportunity to demonstrate the possibility of more sustainable, low-carbon futures and "averting looming ecological catastrophe."[116] This characterization is found frequently across the urban experimentation literature, with projects characterized in terms of their "transformative promise"[117] and "improved urban futures."[118] Urban laboratories are likewise characterized as a break from "development-as-usual"[119] and "conceived for change that is intentionally radical."[120]

At first glance, Miami Beach's experimental raising of miles and miles of roads and its installation of fleets of pumps and seawalls may not seem utopian, in that it does not counter or transform existing social or economic urban relations in the sense just described. Rather than radical change, the goal of this large-scale engineering project is ultimately to manage disruption—both from actual floods or spectacular visions of them—in order to maintain (or increase) current property values, tourism, credit ratings, and insurer confidence. This is an experimental yet conservative mode of governing the present and near-future that employs innovation to manage and preserve what already is. Referring to Figure 7, we can note that the adaptive cycle is also an infinity loop. By embracing and responding to disturbance events—sea rise as well as media and scientific visual and discursive narratives that imagine its impact on the city's markets and infrastructure—it is hoped that this experimental form of back loop urbanization will allow overarching social and economic systems to remain in thresholds of safe operating space. "Safe operating space" here does not refer, as it does for scientists like Johan Rockström, to maintaining earth systems within thresholds that allow human and nonhuman life to flourish.[121] Rather, it refers to the reputational and economic parameters necessary to maintain the city's current social and economic configurations. What is therefore secured in these experiments—which both utilize, respond to, and indeed reject apocalyptic discourse that states there is no alternative—is the infrastructural, political, economic, and imaginal safe operating space for a historically specific, valued way of life—and a resilient image of the city.

"We Don't Want to Be a Guinea Pig"

Of course, this seamless, problem-free vision of secure social and economic systems is hardly the only possible modality and object of back loop urbanization. As Jamie Lorimer argues, in the Anthropocene, the lab in which experiments are launched is neither a pristine or sterile space—the modern notion of experimentation based on testing scientific hypotheses in secluded, artificially controlled spaces—but rather the urban environment within transforming social and ecological conditions.[122] Back loop urbanization is equally reliant on and shaped by the local contexts, histories, and metabolisms in which its experiments are situated. In these open-air, real-time trials, outcomes are unknown. Urban experiments rarely turn out as imagined.[123] Indeed, as geographer James Evans notes, there is a second-order cybernetic quality to urban resilience experiments in that, by dissolving the distinction between field and lab, there is no longer an external observer able to observe the lab as a separate, pristine space.[124] Those who experiment instead become part of the experiment themselves in a situation of "ontological uncertainty."[125]

In this vein, implementation of Miami Beach's infrastructure projects has opened unanticipated situations that have disrupted their implementation. Despite the economic promise and widely proclaimed urgency of resilience, there is far from consensus on the city's street raising, and its approach and priorities have come into conflict with its otherwise-celebrated experimental nature. Such conflicts can be seen in the resident protests and subsequent halting of road elevations on Upper North Bay

Road, an ultra-exclusive street that is or has been home to celebrities including Matt Damon and Jennifer Lopez, as well as high-powered real estate brokers. Expressing her concerns about the impact the redesign would have on her two Upper North Bay Road homes, real estate broker Jean Marie Echemendia-Kouri explained to the *Miami Herald*: "Water has to go down, and if my home is below the road, that's where water will go. . . . We have all the stars, all the captains of industry. I'm not saying any street should be an experiment, but why North Bay Road? . . . We're the highest taxpayers in Miami Beach."[126] Similar protests have been voiced by residents in other Miami Beach neighborhoods facing road elevations, where residents say they now face flooding that never existed before due to their homes being relocated below the raised streets, lack of connectivity between those raised streets and driveways, lack of stormwater management alternatives, and potential decrease in property values. Residents have also argued that the roads that are being elevated do not actually flood but are being targeted based on *projected* sea-level rise alone.

Resident protests and lawsuits—which led the city to pause road elevations for several years and to hire a third-party consultant to review and refine its plans—highlight both economic considerations at the heart of the resilience projects as well as local attitudes toward being subject to their experimental nature. "We will not let Upper North Bay Road become another failed experiment," declared one resident living in the zone where construction has now halted to the *Miami Herald*.[127] "We don't want to be a guinea pig," echoed another resident.[128] "You can't come to North Bay Road, one of the most prestigious streets in Miami Beach, tear up our street and get it wrong," declared developer Glenna Norton, whose 12,400-square-foot mansion on North Bay Road—seven bedrooms, eight bathrooms, elevator, gym, and an infinity pool directly overlooking Biscayne Bay—sold for $12.5 million in 2010 to then Miami Heat power forward Chris Bosh.[129] Notably, while this might at first glance help paint a socioeconomic picture of the neighborhoods targeted for road raising, it is also instructive to observe that this amount pales in comparison to those now found in the city's post-pandemic real estate environment: in 2021 the same property was purchased by the CEO of cryptocurrency startup MoonPay, Ivan Soto-Wright, for $38 million.[130]

Other anticipated effects of "Rising Above" involve the nonhuman environment. Beyond the Instagram images of Miami's aquamarine waters, the city's waterways are frequently under no-swim and no-contact advisories due to high fecal bacteria levels from sewage overflows and street runoff. Scientists and environmentalists have long warned that richly biodiverse Biscayne Bay is approaching a regime shift, and in 2020 hot summer waters, low oxygen levels, and fertilizer and septic tank runoff turned the bay suddenly into a dead zone, blanketing the dazzling turquoise waters that normally form the backdrop to waterfront mansion views with dead fish, shrimp, lobsters, rays, and octopi. In an emergency attempt to inject oxygen back into the bay, fireboats and even the city's stormwater pumps were repurposed to aerate the waters.[131] During their initial installation phase, Miami Beach's stormwater pumps were criticized by residents and scientists for potentially exacerbating the already-endangered bay's problems by funneling street runoff into its waters, with one study suggesting heightened levels of enterococci, bacteria that signal the presence of fecal waste, in the bay at six hundred times the accepted state guidelines amounts for recreational waters (claims that Levine refuted as "defamatory" and "sloppy science").[132] Other unanticipated problems include pump failure during power outages during recent storms, lack of emergency generators, and inability to handle heavy downpours.[133] Finally alongside these, Levine's chief engineer, Bruce Mowry—whom Levine feted in 2016 by designating July 20 "Bruce Mowry Day" to celebrate the engineer's leadership in the resiliency projects—was alleged to have designed and built a $25 million seawall on Indian Creek and removed protected mangroves without environmental permits and U.S. Army Corps of Engineers approval, under emergency declarations and pressure from Levine to avoid red tape and rush the job amid slow approval on federal permits.[134] The allegations led to the freezing of construction work on the seawalls and Mowry's termination in 2018.[135]

These developments suggest the contingency and incoherence common to risk governance that critical scholars frequently emphasize. However, the point of noting the disjuncture between the often flawless imaginaries of resilience and its on-the-ground socio-ecological reverberations is not simply, as geographers Peter Adey and Ben Anderson have pointed out, that governance is not always successful and that its performance and materials can "surprise, shock, enthrall, and disrupt."[136] Relatedly, James Brassett and Nick Vaughan-Williams have called for attention to the contingency of resilience practices, emphasizing their "breakdowns" and "faults."[137] Such qualities may be seen in Miami Beach's story, but these developments also and moreover speak to what is frequently the disjunction between imaginaries and realities of resilience. And while there may be little difference between present conditions and futures imagined in Miami Beach's experiment, in reality the city's actual future will be created in part via the interplay between resiliency interventions and local socio-ecological contexts and legacies. Such is the nature of back loop urbanization: a process taking place amid the uncertainty of Anthropocene contexts, in which resilience efforts are shaped by contradictory, on-the-ground realities as much as projects' own aims and imaginaries. In any case, the $650 million project is only designed to secure Miami Beach for the next several decades. What happens, and what is possible, after 2060 and its projected two feet of sea-level rise remains a space of the fundamentally unknown.

Miami Forever

According to Dalby, "Who gets to decide what kind of life will be lived in what biospheric conditions?" is a key question of the Anthropocene.[138] But who gets to *imagine* what kinds of life will, and can, be lived in the epoch is of great significance also. The possibilities for urban life in the Anthropocene will be sketched out and contested on the terrain of the imaginal, no less than on that of the infrastructural.

Media, scientists, artists, funding entities, governments, and real estate interest organizations have taken up the task of projecting specific visions of life onto Miami's present and future. For these and many other stakeholders, the city has become a living laboratory for trialing imaginaries that will define urban life in the Anthropocene (at least in this city). Climate change ground zero, Wall Street South, capital of cool, and resilient city rising above: each of these urban imaginaries is being sketched onto Miami simultaneously, each in its own way attempting to capture and shape the city's complex social and environmental reality in the early decades of the twenty-first century.

Within this confluence, for many governmental decision makers and business and real estate interest organizations, Miami, and Miami Beach in particular, has also become a living laboratory for developing the techniques—designed to be replicable in other vulnerable cities—necessary to rise above rising seas—current and projected—both in terms of flooded roads and potential economic impacts. And, from an economic and urban governmental perspective, Miami Beach's experiments are, in fact, extremely utopian—just not in the transformative or critical sense normally expected by critical thinkers. After all, what could be more utopian than a city that rises above repeated apocalyptic problematizations of it as the next Atlantis to become the most expensive real estate market in the country, the "capital of cool," and an Aa2-rated top destination for tech, finance, and crypto companies and CEOs? Just as heterotopian climate experiments signal to those who desire sustainability transformations,[139] Miami Beach's infrastructures also beckon from the front lines to a different constituency—that of government, finance, and real estate—offering hope that maintaining property values, credit ratings, and insurer confidence amid actual and imagined sea-level rise is possible.

These utopian efforts might appear as the continuation of the speculative grift that has undergirded Miami since its inception. As detailed by Jason Vuic in *The Swamp Peddlers*, during Florida's 1920s real estate boom, for example—of which Miami was the center—far-away land speculators purchased land en masse, sight unseen. Backed by newspaper ad campaigns, payment plan promotions, and boosterist promotional stunts such as those described in this chapter's introduction, they flipped these frequently waterlogged, seasonally flooded, or wetland lots to retired residents of the Northeast, who were compelled less by the land, which they had rarely seen, than by, as Vuic puts it, "a 'vision' of what they and their property would be, what they would someday be, with no more floors to shine, or buses to drive, in a new home in a new town that (a salesman assured them) was somewhere close to a beach."[140] As described by *The Great Land Hustle* author Morton C. Paulson, "the boom seems almost too bizarre to have happened . . . [but] in retrospect . . . it was one of those strange spasms of speculative madness that occasionally seizes mankind."[141]

Viewed as a continuation of this history, Levine, Gelber, Suarez, and their peers might be read as the city's new "swamp peddlers," albeit for a twenty-first-century world of financial and crypto investment and sea-level rise, a delirious or cynical effort by politicians and developers at one last boom before the floods truly arrive. On this reading, what Miami Beach's infrastructures promise is more accurately that maintaining the "Miami way" of real estate speculation and grift amid sea-level rise and impending submergence is possible.

On the other hand, however, perhaps Miami Beach's resiliency efforts also point to a critical *power* within standard resilience approaches: the willingness to recognize that which one wishes to save from contemporary social and ecological upheavals—imaginal, environmental, and so forth—and to develop the imaginal and infrastructural means to do so. In an era increasingly saturated by apocalyptic climate change imaginaries that proclaim there is no other future but ruination, other ways of seeing are often viewed as naive. At a basic level, there is nothing wrong with back loop experimentation geared toward preserving existing systems deemed worth saving by the parties concerned, and countering imaginaries of crisis with imaginaries of stability and serenity. As Kathryn Yusoff and Jennifer Gabrys point out, while scientific future climate projections "often inadvertently become taken as statements *of the future*, thereby acquiring a solidity and authority that is unmerited (because we have no precedent for climate change under current social conditions)," other future imaginaries generated by other actors might equally emerge to counter such proclamations.[142]

Nevertheless, at the same time it is important to note that many other possibilities exist within back loops, including ones suggested by resilience practitioners themselves. Within socio-ecological systems and sustainability science there are growing calls for conceptual and applied work on *transformation* or *transformative* resilience as a legitimate response to and pathway in the Anthropocene.[143] If what I have called the Anthropocene back loop truly is so upending of urban certainties and grounds, then perhaps moving beyond resilience to consideration of far broader, transformative horizons is the only logical next step, and one that Miami is well positioned to explore. What could urban planning and design become if planners, designers, governments, and even developers bifurcated away from the dominant injunction to build status quo resilience and instead put their skills and energy toward carving out other, daring and transformative pathways? If designers, engineers, and planners truly do what they are often alleged to do—treat cities as their canvas—then why not go beyond street raising and take calls for "living with water" to their fullest conclusion, perhaps by terraforming Miami into a new subtropical Venice, a floating or aquatic city, navigable by water taxis and boats, as so many design and architecture students already dream? Living with water could here become more than more of the same—a means and a call, instead, to profound transformation: bringing water into the city in dramatic new ways, remediating the aquatic ecosystems which have been so polluted, connecting to and regenerating the regional water budget, and experimenting with new aqua-urban ways of life. Such is only one idea. The point is ultimately and simply that, in the Anthropocene, far more is possible.

2
NEW HOUSES FOR A NEW CENTURY
URBAN FUTURES SCENARIO VISIONING AND SOCIAL INFRASTRUCTURE

> Scenarios are . . . the most powerful vehicles . . . for challenging our "mental models" about the world and lifting the "blinders" that limit our creativity and resourcefulness.
> —Peter Schwartz,
> *The Art of the Long View: Planning for the Future in an Uncertain World*

Future on Fire

On a warm, breezy evening during the 2018 Art Basel fair, Miami artist George Sánchez-Calderón installed a large-scale model Levittown home on the sands of Miami Beach.[1] Bookended by ocean to the east and a complex of luxury condos and hotels to the west, the house conjured certain visions of the postwar American dream: home ownership, a better future, and a stable Earth. As the sun sank behind the Faena Hotel, home to Damien Hirst's gold-plated woolly mammoth sculpture, the sky above the water turned purple-pink-yellow. White cloud puffs dotted the horizon. A thirteen-year-old boy—who, with his family, had emigrated from Cuba to Miami to escape the Castro regime—played a piano recital. As waves lapped at the shore, Sánchez-Calderón set the house ablaze. Fire licked at the wood structures, incinerating the facade, then interior beams. Night fell, flames grew thicker, engulfing it until nothing remained of the house but black char and smoke. Through the smoke, lights from the living rooms of luxury condo towers shone vividly.

The burning of the house (Figure 8) was suggestive of many things Anthropocene: blazing wildfires that have reduced homes and towns to ash on America's West Coast, the links between urban development and climate change, the delirium of a speculative art-finance economy firing on all cylinders, but above all, the death of the American dream and the many registers in which this event is now experienced, including not only the utter unaffordability of home ownership but also a viable future to many poor and working-class urban residents, for whom envisioned futures aren't (or weren't ever) what they were once projected to be. In Miami's parched, segregated neighborhoods like Overtown and Little Haiti, the city's Jim Crow history persists in the form of extreme poverty and residential segregation, political marginalization and exclusion.[2] While in the twentieth century Miami was a place of upward mobility and opportunity for some immigrants from the Caribbean and Latin America, for whom the city offered escape from authoritarian regimes, personal harm, or poverty, the window of opportunity seems to be closing ever more today as the structural conditions on which it rested have been largely dismantled.[3]

Today Miami-Dade County is home to the richest and poorest zip codes in the United States; has the second-highest level of income inequality among U.S. cities; and with historically high real estate prices, has been repeatedly ranked as having the largest number of residents ages eighteen to forty living with their parents in any American city.[4] As we saw in the previous chapter, the city's long-standing economic inequality reached dizzying new heights during and after the Covid-19 pandemic. With median household income just above $43,401 and median home price at $589,000 and rising, 78.7 percent of household income is required to own a home in the city.[5] Rents rose 32.12 percent between 2021 and 2022 in the Miami metropolitan area, and increasing numbers of residents struggle to make rent or face

Figure 8. Pre-burn and during-burn views of model home installed by George Sánchez-Calderón on Miami Beach. George Sánchez-Calderón, *How to Win Friends and Influence People*, site-specific installation and performance at Faena Festival held during Art Basel Miami Beach, Miami Beach, Florida, 2018. Courtesy of George Sánchez-Calderón.

evictions as developers tear down older, lower-income housing to build luxury developments.[6] According to RealtyHop, in 2022 Miami became the most expensive housing market in the United States.[7] In the same year, amid overflowing homeless shelters,[8] Miami-Dade County mayor Daniella Levine Cava declared a state of emergency over the housing crisis.[9]

Summing up life in Miami in the second decade of the twenty-first century, curator-writer Gean Moreno writes, "In this city, hope has to go so that its house, like so many others, can be demolished to make room for more shiny towers."[10]

Imagining Urban Futures

Today, many scholars argue that Western liberal societies are facing a crisis of political imagination. As philosopher Chiara Bottici puts it, "in the current world of global governance. Politics seems to have been reduced to simple administration . . . in such a world there seems to be no space for imagination understood as the radical capacity to envisage things differently and construct alternative political projects."[11] Today's political world, she continues, is "deprived of imagination."[12] But current forms of government complicate such claims. After all, if we look at resilience planning, we find everywhere projects devoted to *opening* imagination and explicitly geared toward envisaging things differently.

This emphasis on imagination can be seen well in resilience-building efforts underway across Miami. To generate novel resiliency strategies that address the city's past and present exclusions and socioeconomic inequality, practitioners and critical scholars working in and on Miami widely emphasize the need to redesign urban planning.[13] In contrast to Miami Beach's engineering approach, which as we have seen has been criticized for being too top-down and excluding community input, the Greater Miami and the Beaches resilience plan, *Resilient 305*, highlighted inclusion and social justice as key priorities.[14] In these and other efforts, creative techniques of reimagining, and imaginatively planning for, Miami's possible futures have moved to center stage. Miami is not unique in this regard. In the Anthropocene, dramatic socio-ecological change is seen as both looming and necessary.[15] As old grounds give way, statements proliferate in government, design and planning, critical theory and art, to the effect that the Anthropocene "changes everything," marking a rupture point when now-outdated approaches must be thrown out and imaginative new ideas are needed.[16] This perspective is part of a more widespread "revalorization of imagination" within security as well as the social and life sciences in the name of building resilience.[17] In this context, it is considered crucial that we let go of assumptions that the future will look the same as the present. To move away from top-down, one-size-fits-all modern design approaches and generate novel governance strategies, creative techniques of reimagining urban futures are widely prioritized.

As in numerous other cities, one technique used frequently toward these ends in Miami is urban futures scenario visioning. The futures scenario technique was developed by Herman Kahn at the RAND Corporation in the 1950s for the U.S. military and was recalibrated by Shell in the 1970s to profit in any scenario no matter how disastrous.[18] Today, futures scenarios are widely used by urban designers, academics, developmental consultants, and governments as they seek to break with established ways of thinking, imagine the unknown in an era of complex, indeterminate, distributed risk, and generate innovative ways of anticipating and governing catastrophic events both ephemeral and ongoing (flooding, hurricanes, pandemics, terrorism, etc.).[19] But future scenario visioning is especially common now in cities, where it is employed as a creative, pathbreaking tool for finding innovative ways to govern floods, natural disasters, and extreme heat.[20]

Scenario visioning is welcomed not only as a means of fielding climate change and security threats but also for correcting outdated and unjust top-down modern urban planning approaches. Instead of treating cities like blank landscapes and nature like a resource, designers and planners now argue that cities must be viewed as complex, integrated social-ecological-technical systems. Instead of a single "command and control" solution, resilience planners now seek a mixture of diverse, modular, and interlaced systems-based designs working at multiple sites and scales to reconnect urban fragments.[21] To restitch the pieces of the city and improve social equity, planners now seek to include diverse, formerly excluded communities as key components of coupled social-ecological urban systems in the urban design process.[22] In contrast to seemingly technocratic designs, urban resilience experiments often emphasize social issues and are seen as holding the potential to decrease social inequality—itself frequently listed by entities such as the Rockefeller Foundation as a barrier to resilience—via community involvement and participatory design. One might think here of the many fab labs, makerspaces, or innovation hubs that have become common in cities, open or membership spaces where urban residents are invited to experiment with their own creative design. Here urban climate change experiments draw on post-1960s workplace management techniques, such as open, horizontal modes of organization which encourage innovation and creativity,[23] as well as inclusive participatory citizen science approaches. But they are also distinctly cybernetic. As geographer Kevin Grove argues, key to resilience's "designerly" approach is that no single viewpoint or actor can provide answers. To deal with the complex challenges of climate change in cities, approaches able to fold in and synthesize diverse viewpoints and experiences in order to improve governance processes are needed.[24] For planners, governments, and designers working in Miami, the hope is that futures scenario visioning can address the city's history of exclusion, segregation, and historic lack of civic culture, as well as the imaginative challenges of climate change. By bringing in previously unheard perspectives, the aim is to break out of status quo imaginaries and uncover creative solutions to climate change vulnerability, while at the same time pragmatically addressing the city's socioeconomic inequalities.[25]

The Problem of Scenarios

Futures scenario visioning exercises vary in focus and form.[26] Some focus on imagining future scenarios and rendering these legible to external audiences like policymakers or governments via illustrations, first-person vignettes, or diagrams. Others seek to make uncertain futures actionable by devising strategies to achieve desired futures, whether those involve normative transformation or preempting and preparing for crises. Among these, a widely used approach, found in many academic, governmental, and design workshops, focuses on rethinking urban government and infrastructure. In these planning-oriented exercises, heterogeneous groups of individuals—professors, students, politicians, planners, members of nonprofits, and nonexpert residents—are typically invited to spend a day together. The task: consider the climate-changed city in a specific year—2050, for example—and design improved resilient future scenarios or infrastructural plans for it. Workshops often begin with PowerPoint presentations on urban risk and vulnerability (natural disasters, flooding, extreme heat, socioeconomic inequality, etc.). Participants are then separated into breakout groups. Gathered around tables overlaid with city maps and vulnerability diagrams, they are asked to take a bird's-eye view of the city and envision themselves as powerful planners or rulers hatching out transformative governance plans for its future. Rather than preserve the status quo, many such workshops frequently emphasize the need to break with existing systems and envision a transformed city. This is because, as many scenario planners and resilience practitioners now emphasize, deep, far-reaching transformation is the only way forward.[27]

Many transformation-oriented futures visioning and planning exercises have been held in recent years across Miami, and a variety of transformative visions have been floated in them.[28] Notable in this regard was the Urban Resilience to Extreme Events Sustainability Research Network's (URExSRN) 2019 *Envisioning Future Urban Resilience to Climate Change in the Greater*

Miami Area.[29] In the scenario workshop, which focused on co-developing strategic visions for radically transforming the city's infrastructure for a sustainable, resilient future, participants imagined creating autonomous local food production networks connecting Miami's Redlands agricultural region to Miami Beach community gardens, as a means to break with global industrial food systems; eliminating Big Sugar from South Florida and from the American diet; local neighborhood self-organization; a future where manatees swim freely and no longer bear brutal scars on their backs from yacht traffic. Conversations repeatedly returned to the subject of class, corrupt politicians, and the need for residents to organize themselves rather than wait for governments. Or, in other cases, far from envisioning positive transformative change, visions tended to toward the creatively *dystopian:* floating co-working centers connected by drones and remote work, the eminent domaining of whole neighborhoods to build wetlands, or the need to create a Hydro Authority to police new aquatic territories and enforce hurricane evacuations.

Despite their transformative aims and the undeniably imaginative visions generated by participants, however, as critical scholars and resilience practitioners note, too often, many other urban resilience scenario exercises produce generic or conservative visions.[30] Often by the end of many scenario visioning activities—when participants go home and summary reports are designed—off-the-cuff transformative imaginaries frequently dissipate into generic proposals that mirror existing government projects or already-seen-as-acceptable responses: gray and green infrastructure, "living with water," green economies, flood retrofits, and barriers. Little in them changes the structural life conditions for ordinary urban residents, the texture and fabric of urban life.

Seeing Like a State in the Anthropocene

While scenarios promise the opportunity to envision futures that are actually divergent from the present, the conservative nature of futures scenarios is a problem that critical scholars and resilience practitioners note frequently. For example, sustainability scientist and longtime scenario planner David Iwaniec cautions that "participants may have difficulty engaging in transformative thinking that radically departs from the status quo and challenge existing, dominant systems," a problem that Iwaniec and colleagues are working to overcome via a redesigned futures scenario workshop methodology that explicitly forefronts transformative thinking.[31] For other scholars, such as Claudia Aradau, the problem is not this or that method but scenarios themselves, which they see as inherently neoliberal, fatalistic, or mechanistic.[32] But rarely addressed in discussions regarding the conservative tendencies of futures scenario workshops is the spatial imaginary often built into them, and how, rather than opening a space in which "anything" can be imagined, what scenario workshops actually do, in effect, is enlist participants to see in particular, already-existing *managerial* ways. Positioning them in a sovereign stance, taking a bird's-eye view of the city—which workshops leaders and PowerPoint visuals emphasize must be understood as a complex and vulnerable social-ecological-technological system amenable to governing via diverse, situated infrastructures—scenarios do not open a field to imagine "anything," but instead enlist participants to "see like a state."[33]

As described by James C. Scott in the 1990s, state sight imagines cities, populations, even environments as objects or subjects to be governed—guided and tamed, domesticated and ordered—by sovereign planners. For Scott this way of seeing undergirded diverse twentieth-century catastrophes from colonial development and Soviet collectivization, to factory labor management and the urban planning of Le Corbusier and Robert Moses. In Scott's view, these large-scale, one-size-fits-all efforts to transform societies and cities imagined as their beneficiaries an impoverished abstract subject, standardized and with uniform, interchangeable needs, inhabiting an equally impoverished, uniform, and empty landscape. Viewing their object of intervention thus, governments and planners applied modern instruments of statecraft—censuses, data, mapping, education, media, statistics, security techniques—to build the "dream" of the "rational city," as M. Christine Boyer has described it: legible and ordered, standardized and harmonious, freed from uncertainty and irrationality.[34]

The recent governmental embrace of resilience is, however, radically transforming state sight. In fact, the adoption of resilience frameworks has led to widespread critiques of top-down, engineer-led planning, with designers and planners reenvisioning cities through the lens of socio-ecological systems thinking as sites of complex and precarious human-nonhuman entanglements.[35] "Seeing like a state in the Anthropocene," as we might call this recalibrated optic, incorporates elements of critiques of governmental sight—as in, for example, Scott's own counterposing of local *metis* (practical knowledge and skills) to state sight; political theorist Warren Magnusson's counterposing of city sight to state sight; or geographers Ash Amin and Nigel Thrift's idea of "seeing like a city," in which cities are entangled living networks of human and nonhuman actants—as a means of overcoming the perceived problems of modern urban government.[36] In this vein Amin and Thrift have criticized the "robotic" application of complexity science to urban government, which attempts to order cities "from a vaulting vantage point floating above the fray of the living city unhampered by the nuances of urban process or complex thinking."[37] To remedy this, they call for urban governments to "acknowledge intelligences already at work in the city, address situated problems and set urban governance a challenge of harnessing this plurality rather than subjugating it to a master intelligence."[38] Rather than an opposition to it, this vision of a new urban intelligence mirrors existing practices and forms of sight already deployed in urban resilience government.

This terminology emphasizes an important fact: while resilient urbanism significantly *transforms* modern state sight, it does not in any way *reject* its fundamental aim: governing people and environments. That is, despite being the product of *critique* of past governmental approaches, like modern state sight, the urban resilience optic *remains a mode of government,* whose aim is the production of governable spaces and beings (albeit now seen as complex systems). To do so, resilient urbanism combines the bird's-eye view, hubristic total planning of modern state seeing with Anthropocenic critiques and urgencies. While rejected discursively by urban resilience planners, fundamental elements of modern city planning such as its audacious scope and technical expertise of engineers are also maintained and articulated with critical approaches (entangled networks, bringing nature and marginalized communities in, landscape design, imaginative exercises, etc.).[39] The result is a novel urban security approach well encapsulated in Bjarke Ingels's description of his firm's Rebuild by Design–winning flood defense design for New York City, the Dryline, as "the love-child of Robert Moses and Jane Jacobs."[40] While based in the rejection of modern urban planning's hubristic dreams of mastery (Moses), urban resilience forwards its own hubristic dream of mastery, once more envisioning the planned management of whole cities—albeit this time as volatile social-ecological-technical systems managed via situated, self-organizing infrastructural systems (Jacobs)—and calls for audacious engineering experimentation toward this end (Moses redux).

Anthropocene state sight is thus a way of seeing that both *calls for* and *mobilizes* imagination, while at the same time severely *encaging* it. Rather than opening a space in which one can imagine "anything," too often resilience-oriented futures scenario workshops function as a means of disseminating and practicing this Anthropocene state sight. Rather than the outcome of a free, collective deliberation, in these exercises problems and appropriate solutions—usually, maintaining urban systems and infrastructure, respectively—are defined in advance. Participants must then work within those already-existing parameters to envision future possibilities. As Aradau writes of preparedness exercises, ultimately the "unimaginable" is routed back to tactics able to achieve the generic goals of resilience: "stabilizing identity and retaining the system's main functions."[41] As in the previous chapter, this often means maintaining real estate markets and investor confidence. Here experimentation and innovation in themselves do not necessarily have a relationship to significant transformation—of urban space, life, or futures—but instead recognize and incorporate difference solely as a means of generating new modalities of government.[42]

This is a matter of what Julian Reid describes as the way resilience thinking colonizes and enframes imagination today, but also, more broadly, how governmental thinking does so as well.[43] In addition to seeing *like* government sees, urban futures scenarios participants must also see themselves *as* government. Adopting governmental sight requires abstracting the *city* into a system of systems, in which urban dwellers, natures, and technologies become interchangeable and governable via feedback. It also entails a process of abstraction for *participants*, who by adopting the viewpoint of government are cut off from their own situated realities and the particular desires, contexts, and needs which animate them (which would certainly lead to other understandings of the "problem" as well as appropriate responses).[44] These latter—the actual fabric of each participant's world— vanish as participants imagining themselves as herders, governments, states—the sovereign power that sees and decides for the whole—rather than as strategic actors in their own lives. They are thus separated not just from their own worlds but also from the ability to reject existing conditions and create other ones.

Return to the World

Modern spatial imaginaries and their bird's-eye view have long been criticized by scholars.[45] As geographer Derek Gregory puts it, this way of seeing "disavows its dependence by adopting an objectivist epistemology that separates itself from the picture as an autonomous, all-seeing 'spectatorial' subject."[46] For many critical thinkers, rather than continuing the hubristic, modern logic that scholars like Frederic Neyrat see in resilience efforts, living in the Anthropocene requires a scalar shift: coming back "down to earth," as Bruno Latour has put it, to develop local, situated, context-specific arrangements within messy more-than-human entanglements.[47]

Following this shift, other futures scenario approaches seek to reactivate ordinary urban dwellers' situated viewpoints as well as their capacities. This form of scenario visioning turns away from state sight's "God's-eye view," to use geographer Jen Jack Gieseking's phrase, and toward a form of geographical imagination long described by feminist geographers as always situated and specific to a person and their context.[48] Here we might consider the South Florida Serious Games Disaster Preparedness Exercise, a Category 5 hurricane simulation and community preparedness exercise initiated by a group of Miami nonprofits led by voting and political advocacy organization New Florida Majority and Michael McDonald, director of Global Health Initiatives, who worked in disaster recovery including in Haiti after the 2010 earthquake. Part of the Florida Disaster Resilience Initiative, the daylong disaster simulation held at Miami-Dade College's downtown campus brought together more than one hundred residents from poor and working-class neighborhoods across the city to audit their communities' existing abilities to withstand disasters.[49]

Inspired by self-organized Hurricane Sandy recovery efforts in New York City, the role-playing game used present facts and future trends to design plausible disaster simulations. Held in the summer of 2018, the simulation was designed to generate an atmosphere of emotional and cognitive confusion (as in a real disaster) and, through this, strategic reflection and brainstorming. Participants, who hailed from primarily low-income areas including Opa-Locka, Little Haiti, Overtown, Shorecrest, and Redland, were guided to sit at tables organized by neighborhood or city area. Groups were directed to discuss their reactions to catastrophic events announced one after another over the room's loudspeaker: "It is May 6, 2019. Scientists warn that rate of Antarctic ice melt has tripled in last decade. . . . Turkey Creek Nuclear Plant explosion. . . . Relocated Puerto Ricans. . . ."

Animated, often passionate conversations ensued. With Hurricane Maria's devastation of Puerto Rico still fresh, stories were shared of the nightmares unleashed when infrastructures break down: spoiled prescriptions, contaminated drinking water, empty gas tanks, and no way to transport the sick to hospitals. In these discussions, the high social, political, and economic stakes of Anthropocene futures were clear: one need only consider the entrenched inequalities and class and race structures, and asymmetrical causes and effects of disasters, as they play out in Miami's Black neighborhoods, such as Liberty City and Overtown. After its 1896 incorporation the City of Miami enforced state segregation laws, confining Black residents, who had come from the Caribbean and American South to construct the Flagler railroad, within a fifty-block area northwest of the main business district, called "Colored Town" (today's Overtown).[50] The crowded community was largely unelectrified, lacked plumbing, and was composed of dilapidated, self-built homes.[51] In the late 1930s, tourism and development were on the rise in Miami, and white business and political leaders sought ways to expand the business district.[52] During this time, using unsanitary conditions and containment of disease as a justification, a group of downtown business elites obtained support from the Public Works Administration's Housing Division to build an all-Black housing project, Liberty Square, which would form the center of a new Black neighborhood farther from downtown.[53] While welcomed by Black church leaders and businesspeople as a positive solution to impoverished living conditions, Liberty City originated as an attempt by white (and in some cases, wealthier Black) business interests to create a new Black neighborhood to which Overtown's poor Black population could be relocated, thus making way for private real estate development and downtown expansion.[54] Developers further built a wall up Northwest 12th Street to separate the Black neighborhood from white residents nearby.

From the 1930s to the 1950s, to some working- and middle-class Black Miamians from the West Indies, Caribbean, and

American South, Liberty City represented the possibility of a suburban dream, in the form of home ownership, Black-owned businesses and theaters, safe parks and churches. Similarly, until the 1960s, despite its poverty, Overtown flourished as the "Harlem of the South," a mixed-income cultural mecca that was the destination of iconic African American musicians and intellectuals like Billie Holiday, Louis Armstrong, and Sammy Davis Jr., who, after performing in Jim Crow Miami Beach resorts, came "over town" to sleep and perform after hours in the segregated but vibrant neighborhood's theaters and clubs. However, by the 1960s, historian Marvin Dunn argues, integration allowed many residents to move to other, previously restricted parts of the city, white developers constructed massive rundown rental housing in Liberty City, and the expansion of I-95 and I-395 directly through Overtown displaced 75 percent of the community's population. Many low-income Overtown residents relocated to Liberty City. Alongside the influx of drugs, Liberty City and Overtown were systematically ghettoized in the latter part of the twentieth century, transformed into impoverished places where residents were, as historian N. D. B. Connolly puts it, "trapped in a web of stop-and-frisk policing, insufficient city services, unemployment, price gouging, and worsening rental conditions."[55] Riots broke out in the years that followed, including the 1968 Liberty City riots during the Republication National Convention and the 1980 McDuffie riots that followed the acquittal of the white and Hispanic police officers who murdered Black resident Arthur McDuffie after a traffic stop.

These entwined legacies of segregation, impoverishment, and political marginalization continue to shape Miami's present, with many Black residents living below the poverty line, unemployed and politically disenfranchised from Miami's largely Hispanic and white political establishment, and subject to intensive policing and high crime rates.[56] In 2017, Liberty City's poverty rate was 42.9 percent, and 46.2 percent of households had an annual income less than $20,000.[57] Over half of Overtown families live below the federal poverty line.[58] As Serious Games participants repeatedly recounted in their discussions, fresh food and grocery stores are limited, and food insecurity, already widespread on a non-disaster day, was heightened post–Hurricane Irma when the neighborhood saw food shortages, sweltering heat, and extended power outages. These post-disaster stories were not limited to those from Liberty City but were expressed in varying ways by white, Black, and Hispanic poor and working-class residents of Little Haiti, North Miami Beach, and Miami Beach. Likewise, the memory of disaster continues to be present for residents from the Redlands and Homestead, agricultural communities on the southwestern edge of Miami-Dade County which bore the brunt of Hurricane Andrew's devastation in 1992, a massive Category 5 storm that left 160,000 people homeless.[59] But along with these discussions of the multiple overlapping disasters of inequality, entrenched poverty, and hurricanes, so too were passionate tales of comradery and autonomy from around the city abundant, of commandeering warehouses during Hurricane Irma to cook hotdogs for hundreds of neighbors a day, and fantasies of scaling that up to thousands, ten thousand a day. Other focal points of conversation between participants included the cynical corruption of South Florida politics, favorite preparedness gear, and shared interests in survival skills courses and various localized infrastructures.

As a faux newspaper front page distributed to the tables announced that "A Category 5 Hurricane Is Headed for Miami," participants were directed to discuss what they will do and what capacities they have. Whereas preparedness drills test existing plans, the Serious Games simulation aimed to create a space for communities to gauge their as-is baseline capacities to address "mission critical functions"—communication, flood defense, food, energy, water, and communication, described as organs in a body, prone to failure post-disaster—during disaster events. Rather than taking a bird's-eye view of the city qua social-ecological-technological system to manage, Serious Games participants were to begin from their own neighborhood realities and assess their adaptive capacities or lack thereof ("mission critical gaps"). This scalar shift moved away from "the" city conceived as a total, governable thing—too big and abstract to be a political site, to paraphrase Andy Merrifield—and toward one's own home, body, and world.[60] Here participants left the realm of abstract urban governance, in which they play the role of herders and planners, and came back "down to earth" as theorists like Latour believe the Anthropocene calls us to do, to consider their own realities and needs.[61]

The goal: to build intelligent, agile preparedness networks that augment existing governmental initiatives and entities. After grading their community capacity—extremely vulnerable, resilient, or sustainable—participants moved into "solution teams" in which they were to brainstorm community-led solutions to identified gaps in areas of safety and security, emergency management, communications and education, and infrastructure and basic necessities. Ideas floated included community resilience hubs, charging stations, solar generator stations, food, water, and medicine storage backup facilities. At the workshop's close, teams presented their ideas to a panel of elected leaders, emergency management officials, and resiliency experts. The goal, as an informational flyer related to the scenario put it, was as follows:

> Building a network of citizens with specific subject matter expertise, who can provide assistance locally before large scale disaster relief is available, and can extend and augment regional disaster response in coordination with existing disaster management. Just as the cells of the body each possess the full genetic code, but differentiate as needed to form the interrelated systems of the human body, coordinated by the brain, neighborhood networks and subject matter networks are expanded, connected and enhanced by the Florida Disaster Resilience System.

It is not difficult to see the immediate positive impact such networks can have. Whereas Miami's waterfront mansions largely remain empty during hurricanes, the city's poor often lack cars or means to leave and must ride storms out in their homes. At best, this might be a hurricane party with one's friends and family, packed into small homes with all the extra water and barbecue pizzas the group can gather. At worst, it means shuttered inside unfortified, mold-ridden apartments with mattresses over the windows, little to no food reserves and no generators. There is thus important value here in the raising of one's community up to a higher level of preparedness, and the networks generated at the Serious Games continue to seek powerful political possibilities for low-income Miamians in manifold ways. Furthermore, with Miami now the most expensive housing market in the United States, real estate speculators are increasingly seeking to capture new forms of what Zac Taylor and Manuel Aalbers call "risk rent" by developing the city's poorest neighborhoods, many of which are located at higher elevations and thus less vulnerable to flooding.[62] In fact, the key resiliency issue highlighted by climate justice community coalition Miami Climate Alliance in their 2020 report was not sea-level rise but the crisis of affordable housing in the city.[63]

At the same time, however, especially given the long history of urban schemes that promise improvements to the poor and marginalized while ultimately propping up existing social and economic status quo, one should also remain attentive to the ways in which adaptive capacity and resilience approaches are being taken up and how these can, perhaps inadvertently, *constrain* poor and working-class future prospects to fulfilling neoliberal, biopolitical functions. More specifically, we need to consider how even "earthbound" imaginaries of the potential and limits of the poor's adaptive capacity resonate with recently ascendant governmental concepts, strategies, and imaginaries—and how they might push beyond these.

Social Infrastructure

The naturalization of crisis and disaster and the valorization of vulnerability and survival is central to neoliberal regimes.[64] It is also a common trope of Anthropocene critical thinking.[65] But contemporary resilience thinking pushes further, not just asking urban dwellers to reenvision themselves and their neighbors as surviving machines but seeking to reimagine and harness this survival as key to the *city*. Here we encounter an emergent urban form: "social infrastructure." Used with increasing frequency by urban theorists, planners, and government,[66] this term might bring to mind AbdouMaliq Simone's concept of "people as infrastructure," describing the way the urban poor, often living without robust infrastructural supports, "make use of whatever comes along," "borrow all that is in sight," and in so doing make and remake cities.[67] First elaborated through his work on the inner city of Johannesburg, Simone's insight is that people's myriad, heterogeneous, and seemingly disconnected yet simultaneous actions and practices combine in messy, nonlinear ways to make and remake cities, urban surrounds, and the field of possibility for other actions. These actions and interactions, for Simone, concern the way people make use of urban environments *beyond* survival, as well as, crucially, the creation of a free, open-ended milieu in which the future is not predetermined. Understood from this perspective, these actions and capacities are that very thing which makes life more than bare life; they are that which makes worlds.

But "social infrastructure," in contrast, needs to be understood and situated as a governmental discourse and object produced within the recent urban resilience problematization, in which redefining and recalibrating management techniques on which the city has long relied to administer urban processes has become increasingly synonymous with rethinking infrastructure. Alongside traditional types of critical infrastructure which have been the focus of city and national security agendas since at least September 11, 2001, infrastructure is being redefined to address the perceived problems of modern urban governance based on a separation of city/nature and ideas of control and mastery.[68] Responding to the need to recalibrate modern urban governance and infrastructure, urban and federal governments have begun promoting self-organizing and data-sharing human communities as "social" or "human" infrastructural systems as a supplement to brittle, obdurate technical infrastructures, necessary to urban recovery after disasters.[69] This specific notion of social infrastructure emerged as especially prominent in New York—another global city now seen as a climate resilience lab—within the expert and media narratives that emerged after Hurricane Sandy struck the city in 2012. In that post-storm moment, the then unformed urban resilience dispositif was assembled across media, expert, and government discourse. Just as politicians described the city's technical infrastructure as inadequate for the "new normal," sociologist Eric Klinenberg argued in several public forums and in online media that the city had "impoverished social infrastructure" and called for design experiments to improve it alongside hard infrastructures; critics of municipal resilience projects likewise called for social infrastructures to be given equal consideration and include vulnerable populations.[70] Ad hoc local recovery hubs set up after Sandy, such as Rockaway Beach Surf Club, were championed as resilient nodes of community connectedness; media heralded the swarm intelligence born in New York City's "first social media disaster," noting how the most up-to-date and accurate information was uploaded by social media users rather than traditional news outlets, while the efforts of the activist organization Occupy Sandy were hailed by their unlikely allies the Department of Homeland Security.[71] Now a common trope in many disaster-prone cities, with social infrastructure, the adaptive capacities of individuals long celebrated by neoliberalism are themselves being redefined by cities and academics as infrastructure.[72] This is because as FEMA itself agrees, prepared and connected neighbors add to a city's ability to "bounce back," while isolated and helpless citizens subtract from it.[73] To build up their social infrastructure, municipal institutions in cities across the United States now encourage urban dwellers to see themselves as integral to urban security efforts via a host of emergency preparedness initiatives, including community needs assessments, scenario workshops, community emergency response trainings, and community resilience hubs. "These trained volunteers are a force-multiplier for us," explains Miami Beach Fire Department's operations division chief and emergency manger Juan Mestas, speaking of the city's Community Emergency Response Team (CERT) program, which trains residents, condo associations, and businesses in basic emergency response.[74]

Social infrastructure is a historically specific urban governmental strategy, form, and imaginary that extends what Nikolas Rose has described as the neoliberal governing via (manufactured) community (a strategy often waged simultaneously with the neoliberal fragmentation and destruction of actually existing communities).[75] The contrast with Simone's "people as infrastructure" is useful here. For Simone, the concept "at its core, acknowledges the tyranny of imposing frozen, uni-dimensional categories on messy, evolving social life and social relations that assimilate countless processes and inheritances."[76] But far from producing a field of freedom, "radical openness," or "generating possibilities of acting in concert" as in Simone's people as infrastructure, with social infrastructure, redefining human capacities to communicate with each other, to feed each other or share food and resources, the "grit" to withstand trauma, and so on, are rendered equivalent with smart electrical grids that maintain function during extreme wind gusts, roads that remain passable, or wireless networks maintaining seamless connectivity even during extreme weather events. In this way, human capacities are not valued as world-making techniques—"where the interactions among bodies and materials engenders new constellations of sense and capacity."[77] Nor are they understood as world-transforming techniques, through which people might resist intolerable conditions (or even simply transform urban systems, as in the form of Anthropocene state sight discussed earlier in this chapter). Instead, they are redefined as key components of governing and maintaining other, existing urban systems.

According to Amin, what is often imagined by resilience practitioners as the ingenuity of community crisis responses are, in reality, often activities that the urban poor have no choice but to devise given their lack of infrastructural access and the precarity and exploitations generated by neoliberal regimes.[78] But with social infrastructure, what is considered essential is not only urban dwellers' ability to ride out disasters—governmental abdication of provisioning and administering populations, as critics of resilience qua neoliberalism typically point out—but this ability's effect on maintaining the broader *urban* system. As many critical urban scholars have shown, built urban environments are not simply neutral backdrops. Instead, the roads, energy networks, luxury towers, and so forth that compose urban systems are the concretization of past urbanization processes and, as geographer Jean-Paul Addie points out, the ossified dead labor power of those who built them.[79] In being reimagined as social infrastructure to help preserve and manage these infrastructures, urban dwellers cease to be liberal subjects. Instead, cybernetically they become an augmentation for the broader urban resilience machine—key nodes of adaptive capacity alongside other critical infrastructures. Here resilience's optics function, as Reid has described resilience broadly speaking, as a regime of imagination that circumscribes the possible, "the limits of which are defined by survivability."[80] Whereas the first form of Anthropocene state sight discussed above sought to imagine transformative change, albeit circumscribing this to what can be seen from a governmental view of an abstract infrastructural city, the horizon of transformation for urban communities seen as social infrastructure is becoming better able to endure a disaster-filled future. In this second form of Anthropocene state sight, imagination is directed not outward to govern others or the city qua infrastructural system, but inward toward one's own individual and community life imagined as a biological thing to secure but also a critical element of adaptive capacity in and for an increasingly risk-prone city.

According to anthropologist Bjørn Enge Bertelsen, the city (polis) has long been seen as the site for imagining and trying to create emancipatory transformation. But the resilient city, by operationalizing and thus *capturing* human capacities, and framing resilience as its only horizon, is instead more like what he calls (speaking of African and postcolonial cities) a "post-polis urban context oriented towards controlling bios and zoe and towards precluding the emergence of alternative potential trajectories of development, enrichment, emancipation."[81] Instead of a polis (political space), with all its potentiality—for beauty, strife, encounter, emergence, and transformation—what emerges instead is an anesthetized space, in which these human capacities and qualities are snuffed out. In essence these are insights already glimpsed at the rise of resilience in the early 2010s by thinkers like Reid and Brad Evans, who early on sketched out the key contours of resilience as a regime of imagination and human subjectivity fundamentally oriented around acceptance of fragility and vulnerability and an image of human life reduced to survival.[82]

The potential resonance between the imaginaries embedded in workshops like the Serious Games and those of social infrastructure is clear. But despite the potential capture of human adaptive capacities within a social infrastructural imaginary, can collective consideration and building of these capacities instead be a means of breaking with standard resilience's regime of imagination and opening other imaginaries and material pathways altogether?

Reappropriating Social Infrastructure

In the Anthropocene, cities face at least two crises of the capacity to dwell. The first is a literal material crisis of housing affordability. The second is a crisis of the human capacity to imagine and create worlds beyond those that already exist, a crisis of the ability to image life beyond survival, whether one's own or that of existing urban systems. While these crises can be seen powerfully in Miami, they extend far beyond to urbanized spaces across America and the globe.

Is this all that's left to imagination and to human agency, training to better endure industrial liberal society's disasters and help administer—and become part of—the systems that produce them? Surviving in the charred ruins of the old "house," to return to Sánchez-Calderón's image? But surviving "late capitalist ruins"[83] is what so many urban dwellers already do daily—not a horizon to aspire to. If there is going to be a better future than this—if the capacity to dwell on Earth, to create worlds on and with it, is to be reclaimed—urban dwellers, theorists, designers, planners, and residents alike must find imaginal and material ways to refuse currently dominant imaginaries of urban life in the Anthropocene.

With this in mind, I conclude by returning to futures scenario exercises and the Serious Games. Generating imaginative outcomes that address large transformative questions is the aim of most futures scenario workshops and resilience projects. But too often both produce homogeneous outputs and forms of sight that encage participants' heterogeneous aspirations within existing governmental imaginaries, in which the agency of the working class and poor is limited to enduring inevitable disaster. Here resilience, forwarded as the only response to so-called climate crisis, as Erik Swyngedouw puts it of apolitical climate technosolutionism, "exhausts the horizon of our social and political aspirations and imaginations,"[84] the latter reduced to a biopolitical means of enhancing multiscalar adaptive urban infrastructural capacity. This is a limited imaginary of the city. But it is also a limiting of human subjectivity and the human imaginative capacity itself. As Reid emphasizes, imagination is not just a means of envisioning what already is or "accommodating ourselves to whatever we perceive to be threatening."[85] Imagination is, rather, the human capacity to push beyond the parameters in which one is confined, a means of transgressing, overturning, transcending, or pulverizing them.[86] In this sense, imagination is one art or practice of moving *beyond* bare life, to *become* a subject in the first place.

Despite the frequent limitations of scenarios, collective consideration of potential urban futures—flooded roads, soaring inequality, invasive animals, or drinking-water shortages—can open other pathways in this regard. Scenarios after all are at their best a "space to question the limits of what is normally considered possible, desirable, or inevitable in the face of future challenges," as Iwaniec puts it, and what is clear in many scenario discussions—such as those at the Serious Games—is that dispossession, precarity, and insecurity—learning to stay with the trouble and survive in ruins—are not seen by poor and working-class urban dwellers as something to be embraced but neoliberal impositions to be challenged.[87] Toward this end, a conclusion drawn repeatedly in these and other discussions of Anthropocene urban futures, such as those of the Serious Games, is that, to resist neoliberal-imposed socio-ecological precarity and looming austerity, there is a need to engage materially and pragmatically with transforming environments by learning old and new survival skills for the urban Anthropocene. Indeed, for many urban dwellers, learning "survival" skills represents not just a means of ensuring their own or existing urban systems' continuation, but instead means of building alternative urban futures or ways of being with their own hands. The uses and exploration of solar generators, goTenna Mesh devices—which enable users to create pop-up wireless and send private text messages—or the rescue feats of the Cajun Navy along the hurricane-battered Gulf Coast are common conversational topics in urban futures scenario workshops. As are shared musings on their potential expansion beyond individual hobbies.

This interest mirrors a broader, multiracial, multi-aged, multigendered phenomenon: the growing normalization of once-outlier activities among increasing numbers of individuals, families, and communities across urbanized America. In Miami one finds a small but growing number of urban farms and food networks, such as Little River Cooperative, an urban farm and CSA that also seeks through workshops and classes to, as they put it, "empower people to grow their own food and medicine at home."[88] In a similar vein, as geographer Melissa Bernardo shows, the multidimensional crises of recent years have inspired the emergence of manifold food-production networks of solidarity across Miami-Dade, particularly in the Redland agricultural zone.[89] And on the tech side of things, to meet their strategic aim of making the predominantly Black municipality of Miami Gardens a sustainable city able to "leapfrog the supply chain" using end-to-end economy, in 2015 EcoTech Visions Foundation opened a 52,000-square-foot creative coworking, manufacturing, and incubator space. Aimed at democratizing and activating capacities that would normally "remain dormant due to limited resource availability," along with space and ecofabrication equipment and a FarmTech growlab for future food production, EcoTech Visions provides free coding, solar, technology robotics, digital fabrication, "green scout" boot camp classes, and fellowships to communities who would not normally be able to afford them.[90] Alongside these survival skills and tech, given the fact that many Miamians view the housing crisis as a far more pressing threat to their survival than climate change, strategies for staying in place are also widely seen in Miami as Anthropocene urban survival skills themselves.[91]

Extending far beyond Miami, the learning of past and present survival practices such as how to make a fire, field dress and quarter a deer, bow hunt, track, navigate by stars, or prepare for blackouts has never been more popular, and in the wake of recent hurricanes, supply-chain crises, and the Covid-19 pandemic, bushcraft, survival, preparedness, and wilderness skills classes are at capacity across South Florida and indeed across America. In backyards, on fire escapes, or via territorial networks, many inhabitants of cities, suburbs, and hinterlands across the United States have begun constructing high- and low-tech localized infrastructures designed to reduce dependency on the volatile systems of planetary urbanization: wireless mesh networks, rainwater harvesting systems, protein bar engineering labs, post-petroleum resilient homesteads, or a "civilization starter kit"—open-source blueprints for the fifty machines needed to rebuild one's own civilization. In repeatedly flooded areas, others people are designing amphibious architecture and trialing ways to desalinate a contaminated urban drinking water supply.[92]

Of course, one could easily read the orientation toward autonomy in these popular activities as extreme forms of the neoliberalization and biopoliticization of society: a dystopian panorama of humanity at its end, stockpiling solar generators and satellite phones to survive disaster alone at home, carefully shielded from each other and the world outside. Certainly, the production of docile populations, suspended perpetually in crisis, afraid of everything and everyone, is, like social infrastructure, a governmental use of preppers and survivalists. But this turn toward acquiring survival skills and building independent infrastructures when faced with looming urban futures also invites other interpretations. Some of these attempts are motivated, at least nominally, by pragmatic and material considerations related to disaster preparedness: how to inhabit perpetually flooded cities like 'Miami, obtain clean drinking water if salt water infiltrates urban aquifers? Or, as in southern Louisiana, how to remain living in areas now slammed repeatedly by hurricanes (especially when governments and insurers decide such areas have become "unlivable")? But they are also driven by political and philosophical questions that have become central in the age of climate change: how to care for oneself and one's communities? How, concretely, to break away from existing, ruinous socioeconomic systems and take back power over one's life and abilities?

As Grove argues, resilience design practices, including futures scenarios, target and reroute existing, and potentially more potent, socio-ecological adaptive capacities, making particular worlds present by recasting environments as complex interconnected sites wracked by danger and vulnerability in need of and amenable to government. In doing so, he writes, they "demand social and ecological difference present itself in ways that are useful for widespread efforts to synthesize sustainability solutions," or, we can add, administer existing urban systems more broadly.[93] Here, resilience practice too often entails "capturing and ensnaring difference," in order to produce the same.[94] But this operation can be turned on its head. Instead of accepting hegemonic understandings of the Anthropocene as an age of inevitable risk and risk management to secure urban economies, infrastructure, and existing social miseries, "social infrastructural" powers can be directed toward other goals and other futures.

Indeed, rather than a neoliberal dystopia, perhaps the landscape that could come into view as one envisions urban dwellers of all kinds taking stock of current crises and promised futures and, in response, learning to grow their own food, access their own drinking water, 3D-print their own houses, and hunt game—developing their own infrastructure—is more that of populations gathering the preliminary means necessary to inhabit the Anthropocene on their own terms, rather than the ones imposed on them. And the means to do so even in places being deemed uninhabitable. Exploring the pathways this reclamation of imaginal and technical powers opens offers a point of departure for much more far-reaching, larger-scale endeavors and terrains. It also opens imagination to consider much broader horizons. As the old order is already incinerating itself, instead of managing it on life support, it could be that within these trends is the nascent possibility of moving beyond mere resilience and toward envisioning and actually building new "houses"—spaces and modes of dwelling—for a new century.

3
FLOWS AGAINST FLOWS
THE EVERGLADES AS CRITICAL URBAN INFRASTRUCTURE

> Urbanization now intervenes as a program that can be understood only as the *immunization* of nature in which the technology of the urban acts as the vaccine.
> —Ross Exo Adams,
> "Natura Urbans, Natura Urbanata"

From Swamp to Salvation

Urban vulnerability to sea-level rise is usually associated with flooding. In this vein, Miami's climate change endgame is frequently imagined in terms of Zaha Hadid skyscrapers and art deco hotels drowned by glittering blue seas, hundreds of billions of dollars in real estate submerged by century's end. But many scientists and planners predict that the city's more urgent, and imminent, sea-rise threat is not flooding or land loss but saltwater contamination of the Biscayne Aquifer. A permeable, limestone strata that lies beneath Miami, the surficial aquifer supplies nearly all of the freshwater used daily by residents, businesses, and agricultural industries across South Florida (Figure 9). In recent years, scientists, water managers, and media have begun warning that, as Earth's oceans warm and ice melts, a wall of Atlantic Ocean salt water is advancing inland, seeping into the aquifer through canals and pushing westward underground—"a sort of Sherman's March in reverse, as prosecuted by the sea."[1] The "Anthropocene marine transgression,"[2] as scientists Randall Parkinson and John Meeder refer to it, has already begun contaminating Miami's water supply, forcing managers to abandon and relocate wells to the western edge of the urban boundary.[3] Rather than floods or hurricanes, according to geologist Harold Wanless, this rapidly advancing salt front may be the real city-killer. In his estimation, if nothing is done to prevent the city's tap water from becoming undrinkable, sewage systems from becoming nonfunctional, then Miami residents will be sent fleeing en masse like twenty-first-century Dust Bowl refugees.[4] Buildings can always be elevated, infrastructures stormproofed, but without water to drink, there is no urban life.

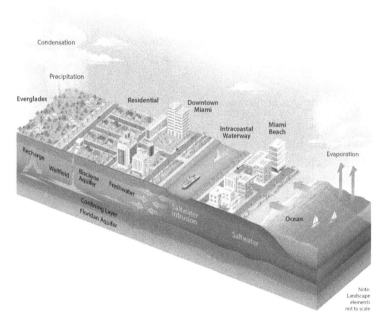

Figure 9. Diagram showing the water systems that subtend and surround Miami, including the Biscayne Aquifer, which supplies most of the city's drinking water and is recharged by freshwater from the Everglades, as well as saltwater intrusion pushing into the aquifer from the Atlantic Ocean to the west. Copyright Hiram Henriquez, 2023.

Figure 10. Aerial view of the Everglades backcountry. Credit: National Park Service, 2020.

This unique existential urgency has generated a novel infrastructural solution, one that shifts our focus away from the city as it is traditionally understood and into the nearby Everglades, a vast subtropical wetland ecosystem adjacent to Miami and that once blanketed much of South Florida in slow-moving freshwater (Figure 10). Long beloved by ecologists, Native American inhabitants, frontiers people, smugglers, and all those "who have felt the strange allure of hidden and remote places,"[5] today the iconic "river of grass"[6] is a paradigmatic example of Anthropocene socio-ecological crisis—pushed to the brink of collapse by modern management, its flows largely interrupted or drained, its well-known alligators, panthers, and wood storks on the edge of extinction. But planners and ecologists believe that restoring the ecosystem's natural water flows is the best hope of saving the ecosystem, and, it is increasingly thought, the city as well. Toward these ends, the South Florida Water Management

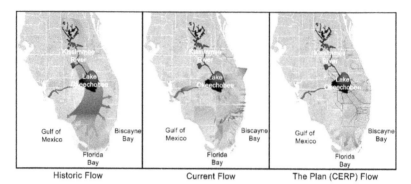

Figure 11. Diagram of Everglades hydrologic flows pre-compartmentalization and drainage, post-twentieth-century compartmentalization and drainage and into the present day, and in the CERP's anticipated future after restoration. Credit: Courtesy of the U.S. Army Corps of Engineers.

District and the U.S. Army Corps of Engineers are engaged in a more than thirty-five-year long, $23 billion experimental effort to restore surface-water flows to their historical, pre-drainage dimensions—or at least something close to them. This Comprehensive Everglades Restoration Plan (CERP), a "pharaonic,"[7] 18,000-square-mile engineering project—the largest ecosystem restoration in the world—is being conducted via experimental techniques in a vast (and vastly altered) landscape replete with technological and industrial legacies (Figure 11).[8] Along with ecological restoration, many scientists and water managers hope that by "get[ting] the water right" in terms of "quantity, quality, timing, and distribution,"[9] the weight and pressure of freshwater will push back against the "intruding"[10] seas, securing urban drinking water supply and, with it, the continued viability of human life in Miami (see Figure 9). Making the city resilient is now seen as a matter of restoring nature's flows outside the city.

Conceived this way, the Everglades are being drawn into the problematization of urban government in the Anthropocene and the associated search for new infrastructures capable of both overcoming modern planning approaches and administering the effects of climate change on urban life. In the preceding chapters, such urban resilience infrastructural experiments were predominantly human. In chapter 1, with "Miami Beach Rising Above" and its industrial pumps, elevated streets, and seawalls, we saw experimental yet still technical infrastructures created *by* humans. And in chapter 2, we also saw the embrace of so-called social infrastructure, in the form of prepared and adaptive human communities and networks—humans *as* infrastructure. Designers and governments view both of these as vital infrastructural elements of a resilient city able to field and absorb the effects of climate change. However, this turn to the Everglades brings us to a third form of infrastructure, in which nature's perceived natural functions and flows are revalued as critical urban risk-management systems necessary to secure urban life.

This is a new imaginary of the Everglades and its relationship to the city, one that pushes beyond recent sustainability framings of the ecosystem. While Calusa, Seminole, and Miccosukee Indians and white glades hunters long inhabited the watery land of sawgrass and reptiles,[11] for nineteenth- and twentieth-century planners and government the wetlands were viewed as an impediment to urban development, profit, and productivity; as anthropologist Laura Ogden sums it up, "miasmic and dangerous, uncivilized, and certainly worthless."[12] To urbanize South Florida, millionaire investors and the state of Florida built large-scale drainage projects to reclaim the "derelict landscape"[13] for commercial agriculture and real estate development, thus creating the ground on which much of the urbanized area was built (Figure 12). Following ruinous hurricanes and floods, in the mid-twentieth century the Army Corps of Engineers created a colossal hydraulic infrastructural control system that channeled and contained this ecosystem's water flows, protecting urban and agricultural areas from flooding but producing catastrophic ecological effects such as degraded water quality and widespread loss of animal habitat. By the end of the twentieth century, however, the damaging effects of this command-and-control management were clear, and scientists, engineers, and water managers had come to see the Everglades as an integrated socio-ecological system in which sustainability could be achieved through the proper adaptive management of its social, environmental, and economic domains.[14] In 2000, Congress approved CERP to reverse the effects of twentieth-century management, restore surface-water flows to an approximation of their historical, pre-drainage dimensions, improve water quality, and restore habitat for endangered species. The result of decades of environmentalist and Indigenous demands to address ecosystem degradation, Everglades restoration was described as a model for successfully undoing twentieth-century environmental damage. Thus at the dawn of the twenty-first century, Everglades management was a vision of sustainability, recovery, and progress, with the largest environmental restoration effort in history described by Jeb Bush as "the restoration of a treasure for our country."[15] Water quality would be improved. Crocodile nests would again crowd the ecosystem's banks and beaches. Roseate spoonbills would return to block out the skies with their brilliant pink plumes.

But as attention to sea-level rise and saltwater intrusion has increased, this vision of sustainability has changed dramati-

Figure 12. A Seminole woman poling past a billboard advertising the sale of reclaimed land in the Everglades, circa 1920. Courtesy of and caption by HistoryMiami Museum.

cally. Transforming earlier visions of harmonious ecosystem restoration—when CERP was founded its scope and calculations included no mention of sea-rise risk—local media, environmental activists, water managers, and policymakers now situate the Everglades as a critical bulwark against advancing saltwater encroachment, an existential necessity for a city threatened by climate change intrusions. Their hope is that restoring surface-water flows to their historical dimensions, replenishing the hydrology by filling in artificial canals and removing levees, and constructing additional storage reservoirs to restock water during climate change-increased dry seasons, will release the regional ecosystem to act as a downward counter-force to push the advancing front of salt water back and secure urban drinking water supply, or as NPR put it, "repel the invading sea."[16] Alongside visions of flourishing alligators and flamingos, planners, politicians, and media now forefront a restored Everglades as the "antidote to sea level rise,"[17] with the capacity to push back salt water intrusion and keep drinking water flowing from urban taps.[18] As James Cason, Republican mayor of the city of Coral Gables, explains his constituents' concern with restoration: "It's not just so they can see the alligators. It's because they'll want to make sure the drinking water on which we all depend is not contaminated."[19] As discussed in chapter 1, resilience in Miami is tightly coupled to preservation of urban real estate markets and investor confidence, with the ability to govern sea rise a critical signal to banks, developers, and credit rating agencies that the city is still viable financially. Thus Everglades restoration is a material but also urban development and economic issue; as emphasizes Miami-Dade chief resilience officer James Murley, "the minute the world thinks your water supply is in danger, you've got a problem."[20]

Anthropocenic Infrastructural Nature

As the Everglades are being enrolled into the urban Anthropocene problematization, the very qualities for which they were once disparaged—flowing wetlands that impeded urban development—are now seen as critical to urban resilience. This is a novel imaginary and valuation of the Everglades' "liquid landscape," which, while reviled by urban developers, has long been valued in manifold ways by diverse inhabitants who called and continue to call it home. In the nineteenth century, the Everglades provided the Seminole and Miccosukee Indians a means of evading capture and the forced relocation during the Indian Wars, a valuation centered on autonomy which continues today.[21] For urban water managers and governments, the Everglades have long been valued as the source of urban drinking water, like so many other forms of what geographer Matthew Gandy calls "metropolitan natures" regarded as a "standing reserve" to provision cities.[22] In this sense, the Everglades can be and have been understood for some time as what Neil Brenner calls an "operational landscape," that is, spaces and territories typically seen as "outside" of cities that are harnessed and engineered to serve the latter's needs: "zones of resource extraction, agro-industrial enclosure, logistics and communications infrastructure, tourism and waste disposal, which often traverse peripheral, remote and apparently 'rural' or 'natural' locations."[23]

In contrast, however, an Everglades restored to pre-drainage flows is today also increasingly imagined and valued as something distinct: a means of managing other, nonnatural water flows—sea-level rise and saltwater intrusion—and the threats they pose to urban infrastructure. Here water is imagined as valuable neither for its own sake—for example, as an ecological treasure to be conserved or as a place of meaning—nor for the sustenance it supplies to the city, but rather for its ability to move with gravity slowly and heavily toward Florida's coasts, and how this can affect and govern the impacts of other, undesired water on urban life.

This reimagining of the Everglades brings the ecosystem's waterways into a new technical realm with which they have never before been identified: critical urban infrastructure. Defined by anthropologist Ashley Carse as "the vast, complex, and changing systems that support modern societies and economies,"[24] infrastructure typically refers to human-made technical structures such as dams, bridges, highways, electrical grids, military installations, and computer networks—invisible yet vital lifelines for industrial liberal urban orders.[25] Infrastructure plays a key role in securing urban bodies, materials, and processes, providing the logistical undergirding of modern cities: roads to circulate commodities or pipelines to manage flows of drinking water and fossil fuels, to name a few examples.[26] Not just neutral but also political, critical infrastructure is a biopolitical matter of producing and reproducing liberal capitalist life, ensuring that the "vitality" and "productive powers" of human life are "available to state and capital," as geographer Bruce Braun has put it.[27]

In the nineteenth and twentieth centuries, infrastructures were a key aspect of modern urbanism—*the* central trope according to architectural historian Swati Chattopadhyay—with large-scale engineering works providing evidence of humanity's power to order external, inert nature into productive flows for commerce and improve life quality.[28] Here we might think of the enormous water management system built in the twentieth century to control the Everglades itself, a complex industrial network of dams, one thousand miles of levees and canals, and reservoirs, and pumps to protect urban and agricultural land from flooding. But in the Anthropocene, the ordering "promise" of modern infrastructure is now seen by many scientists, designers, and policymakers alike as a destructive anthropocentric idea causing climate change and network fragility, among other crises.[29] Modern infrastructures with their city/nature binaries are now cast as outdated, the source of the problem, and urban security is dependent on overcoming them. In the Anthropocene, this "engineer" model, as urban planner Kara Elliott-Ortega characterizes it,[30] is relentlessly critiqued for its siloed or one-size-fits-all plans implemented without regard for nature. As discussed in the introduction, this critique of modern urban planning is a key component of the new problematization of urban government in the Anthropocene, with the epoch of the human interpreted by media and government in global cities from New York to London as introducing not only new risks but also a "new normal" in which modern urban governance based on a city/nature separation and ideas of mastery and control no longer suffice.[31]

In this same register, twentieth-century water management in the Everglades is almost universally understood as a poster child for the devastating failures of the modern approach and its hubristic "schoolboy logic," in the words of Everglades activist-writer Marjory Stoneman Douglas.[32] For resilience pioneers C. S. Holling and Lance Gunderson, twentieth-century Everglades management is a paradigmatic case study of the outdated and "pathological" modern command-and-control scientific management approach, which by decoupling social and ecological systems locked the ecosystem into a change-resistant "management trap" and produced a profound ecological crisis.[33] Not limited to environmentalists and ecologists, a wide range of Floridians too decry how the Everglades hydraulic management has transformed the region's lakes and rivers into polluted, deoxygenated cesspools, as evidenced in the repeated toxic algae bloom outbreaks that devastated coastal waterways in recent years, affecting fishing industries and sending dead manatees, sea turtles, and massive fish kills washing up along South Florida coasts.

The persistence of modern engineering approaches in contemporary urban resilience projects is also frequently criticized. For example, while Miami Beach's seawalls, pumps, and elevated roads, explored in chapter 1, are praised for their efficacy, as part of the growing "resilience"/"we need new infrastructure" discourse, such designs are also frequently criticized as inadequate and outdated. In its evaluation of the "Miami Beach Rising Above" adaptation experiments, the real estate organization Urban Land Institute (ULI) recommended that instead of building only gray infrastructures that hide water underground, the city should take cues from "living with water" approaches under development in Venice and post-Katrina New Orleans to work with nature's flows via new green and blue infrastructures.[34] Instead of, or complementing, twentieth-century "concrete and steel" one-size-fits-all urban plans that block out nature, ULI

Figure 13. Visualization by Curtis + Rogers Design Studio for the Miami Downtown Development Authority of the impacts of the Army Corps of Engineers' Back Bay Coastal Storm Risk Management Study, which proposed building floodwalls and floodgates around Biscayne Bay, blocking downtown Miami from the waters that define the city. Highlighting the negative impacts of gray infrastructural solutions, these visualizations were tied to counterproposals for nature-based infrastructural solutions. Credit: Rendering by Curtis + Rogers Design Studio, Inc., 2021.

recommended blue and green infrastructural approaches that welcome and integrate nature—and especially water—both aesthetically and functionally, and the development of plans for "living with water" not just as an inevitable fate, but an opportunity to develop new forms of tourism and place-making. Similarly, the need to push urban resilience beyond gray infrastructure is emphasized by many across Miami-Dade County, as seen in recent calls for green infrastructure tied to the county's rejection of the Army Corps of Engineers' $4.6 billion proposal to build floodwalls and floodgates around Biscayne Bay, thus blocking downtown Miami from its aquatic environment (Figure 13).

As planners and governments search for ways to overcome modern urban governance approaches, there has been a broad shift toward the idea of using nature as critical urban infrastructure. "Urbanized nature," as Hillary Angelo puts it, of course has long been seen as "good"—socially beneficial—and parks and green spaces have been promoted as solutions to urban congestion and social conflict since the nineteenth century.[35] But the explicit notion of nature as *infrastructure* came into usage more recently. In a 1984 report of the United Nations Education, Scientific, and Cultural Organization's (UNESCO) Man and the Biosphere program, the term "ecological infrastructure" (EI) was forwarded as one of five key principles of ecological urban planning.[36] Rooted in biological conservation and landscape ecology,[37] EI most often referred to the ecological conditions necessary for or conducive to a species' survival or movement across territory[38] and referred to natural landscape features such as watersheds, lakes, or hedgerows.[39] The related idea that ecosystems provide services and embody natural capital has also gained traction in recent years.[40] At stake in this new form of "neoliberal nature"[41] is the commodification, quantification, and monetization of natural systems—as in recent efforts by scientists to calculate the economic value of carbon stored in Everglades mangroves.[42] And as geographers Sara Nelson and Patrick Bigger have shown, making ecosystems "investable" also involves a wider range of political-economic mechanisms and labors through which ecosystems are ontologically "repackaged" as "infrastructural nature" in order to extract value, "unlock" new sources of profit, and generate returns on ecosystem restoration.[43]

However, as urban climate risk has become a key governance concern in recent years, a new thinking of infrastructural nature has emerged which must be distinguished from these neoliberal, economic valuations. Defined in 2015 by Nature Conservancy scientist Robert McDonald as "any piece of nature that provides important benefits to those in a city,"[44] ecological infrastructure is increasingly shifting away from conservation and biodiversity focuses and into the realm of disaster management and urban security as cities seek new socio-ecological designs to govern climate change and build resilience via integrated complex systems models.[45] As we saw in the book's introduction, instead of a single "command and control" solution to nature's incursions—massive sea gates around coastal cities, for example—urban resilience proponents now call for a mixture of diverse, modular, and interlaced systems-based designs working at multiple sites and scales to reconnect urban fragments.[46] Instead of separating the urban from nature, planners now seek ways to bring nature back into cities, as part of a shift away from seeing the latter as exclusively human places and instead as coupled socio-ecological systems.

This urban resilience approach in many ways parallels and draws upon critiques that have existed in urban planning since the 1960s. We might think of architect Christopher Alexander's 1965 critique of "artificial cities" designed like trees with separated systems and his attendant proposal that the natural urban structure is instead a complex semilattice of overlapping, changing living systems.[47] Likewise, in 1969 landscape architect Ian McHarg called for urban architecture to "abandon the simplicity of separation and give unity its due. Let us abandon the self-mutilation which has been our way and give expression to the potential harmony of man-nature" by designing with nature.[48] And in the same decade, many such as Jane Jacobs criticized modern engineering-based urban planning's emblematic developer, Robert Moses, arguing against top-down, one-size-fits-all design and seeing cities as made of siloed components, and advocated instead seeing and designing cities as dynamic self-organizing webs of human and nonhumans.[49]

However, whereas past urban planning critics often advocated entangled human-environment systems approaches as a means to improved life quality and a more harmonious urban, Anthropocene promotion of complex systems thinking is often explicitly couched in a framework of inevitable risk, vulnerability, and crisis. Entangled urban systems are now seen as vulnerable to environmental risk, with rising seas and flooding threatening lives and disruption to urban infrastructures, such as transportation systems, energy networks, or drinking water supply. But interconnected urban systems are also highlighted because using complex systems approaches that include nature are seen as necessary to manage those threats. Hard walls to block out water, for example, are seen as problematic not only because they degrade urban life but also because they represent a failed management strategy. Emphasis is placed on the need to abandon views of human-natural urban systems as linear and built within a stable world, and instead see both as complex, nonlinear, and unpredictable across space and time. Instead of seeking equilibrium or sustainability, urban resiliency approaches stress the need for embracing a new normal of disequilibrium and chronic disturbance.

Enrolled into this context as critical urban infrastructure, Anthropocene natures like the Everglades are imagined and deployed in ways that must be distinguished from the varied forms of neoliberal economic calculation, commodification and financialization by which nature and its "services" are brought into view and mobilized. While the Everglades is also seen and valued as what we might name *neoliberal infrastructural nature*,[50] the newer problematization of urban government into which the Everglades are being enrolled entails an imaginary of it as what I call *Anthropocenic infrastructural nature*. This is a specific conjunctural, historically specific imaginary of nature as infrastructure that draws on but transforms even recent ideas of nature as infrastructure in cities. Here, the natural processes and capacities of living and nonliving natures are forwarded as emblematic of a resilience approach that has been recalibrated for urban government in the age of climate change. As such, waterways and animal life are enrolled alongside other resilience technologies to buffer extreme climate change effects and preserve urban orders. Alongside the Everglades, here we might think of how the imagined natural life cycles of oysters are being enrolled to manage storm surge and flooding along New York's coastline.[51] In this novel class of infrastructure, living and nonliving natures are valued not for their "natural capital"[52] or for their ability to provision cities with critical resources like drinking water or sustenance, but because by "doing their nature thing" they are seen as having the ability to govern other natural processes, namely, saltwater intrusion in the case of the Everglades, so as to secure urban infrastructure. Valued here is not *merely* nature qua complex adaptive system[53] but nature as complex adaptive system that is *also* vitally connected with urban systems and agglomerations. In and through this interconnection of complex adaptive urban and nonurban systems, it is hoped that Anthropocene risks to the former, while never stopped or even mitigated, will be administered by the latter.

Rather than an apolitical solution to modern human/nature binaries (as it is often portrayed), broadly speaking this new idea of infrastructure arose in the specific historical context of the recent Anthropocene problematization of modern urban infrastructure and search for new infrastructures to govern climate change and secure existing forms of urban life. Nature as infrastructure is thus *biopolitical*.[54] Rethought as critical urban infrastructure, nature is being enrolled to support and secure the conditions for a specific version of human life. This, many scholars have shown, is the biopolitical role of critical infrastructures in liberal capitalist societies: to, in Braun's words, reproduce liberal life and undergird the "vitality" and "productive powers of populations."[55] As previous modern biopolitical techniques have come under criticism as the source of contemporary crises like climate change, nature as critical infrastructure, like other ecotechno innovations, is forwarded as a new approach appropriate to the Anthropocene "new normal." Importantly, these projects do not replace past approaches or change existing socioeconomic conditions. Instead, they protect and maintain them by anticipating and managing environmental crises in new ecological ways.[56] Viewed in this context, Anthropocenic infrastructural nature must be understood as a mode of biopolitical governance for new environmental risks, part of a more widespread reconfiguration of governmental technologies to maintain urban systems.[57]

Bringing nature into the realm of urban risk management involves a new imaginary of nature, what it is and does, as well as how it works and must be managed. And while Anthropocenic infrastructural nature transforms the representation and production of urban nature,[58] it is also a new definition of critical infrastructure. In place of—or in tandem with—concrete and steel public works, nature, it is imagined, will perform critical urban risk-management functions by doing what it is imagined to do naturally—and perhaps more effectively than humans or gray infrastructure itself can. To better understand the novelty of this new form of urban critical infrastructure and to flesh out a conceptual understanding of it, it is useful to delve back in to how it works.

Flows against Flows

While tourists usually visit Everglades National Park, that is only a small part of the vast 18,000-square-mile Everglades ecosystem which, along with its iconic sawgrass prairies, is a massive ecosystem that includes cypress swamp forests, rivers, lakes, estuaries, and hammocks. The Everglades is part of an extensive regional water drainage basin, whose fresh waters flow from the Kissimmee River into Lake Okeechobee, filtering through sawgrass marsh and into the Biscayne Aquifer, finally emptying into Florida Bay.[59] These flows are part of a unique hydrogeological landscape. Because southern Florida is largely flat and gently slopes southward, historically Everglades freshwater moved down a hydraulic gradient, filling canals, streams, and wells as it moved slowly down into Florida Bay, Biscayne Bay, and the Gulf of Mexico.[60] Before modern compartmentalization interrupted these flows, an estimated 650 billion gallons of freshwater flowed from the Kissimmee River and into Lake Okeechobee. In summer wet months, water would overflow the lake's southern rim and drift through sawgrass marsh, and over the region's limestone ground in wide, slow-moving sheets.[61] As this freshwater moved south across Florida, it descended from the higher elevations at the state's center to lower-lying coasts and into coastal waters But the waters also seeped down, through peat soil and into the limestone strata, joining with rainfall to fill it with freshwater.

Between the mid-nineteenth and mid-twentieth centuries, much of the "liquid landscape"[62] was drained, dried, and rerouted to facilitate real estate and agriculture, control flooding, and create the ground on which many parts of the region's urbanized areas were built. Today a massive system of levees and canals diverts water away from agricultural and urban areas and toward coastal estuaries, resulting in a 70 percent reduction of freshwater flowing southward through the Everglades.[63] The dense but narrow urbanized zone along southeast Florida's coast is steadily encroaching further into the Everglades, which while still vast is a husk of its former self. The Anthropocenic infrastructural view of the Everglades now seeks to *revitalize* and *restore* this flowing wetland and to harness this perceived natural hydrology necessary to secure the city against unnatural sea rise. As envisioned by hydrologists and water managers, once reconnected and replenished, Everglades freshwater will again flow across the surface and underground toward the coasts and in the process push back against the advancing saltwater front, repelling the freshwater-saltwater interface back and away east and away from the city's drinking water supply. "Putting more freshwater back into the sponge that is the Everglades could create a kind of 'back pressure' to keep seawater out," explains Julie Hill-Gabriel, vice president for water conservation at the National Audubon Society.[64] Thus the natural flows of water are no longer seen as an impediment but rather an existential necessity for the survival of the city.

Restoration advocates often depict rehydrating the once-maligned ecosystem as a matter of returning the Everglades to its natural, pre-engineered state—"as nature intended."[65] In reality the effort entails a massive engineering project described by former Florida governor and U.S. senator Bob Graham as "open heart surgery" for a critically ill patient.[66] Overseen by the same entities that handled twentieth-century drainage and management of the wetlands—the Army Corps of Engineers and South Florida Water Management District—the "re-plumbing" plan involves removing impediments to sheet flow and constructing new infrastructures to store, supply, and time water input and flow. The National Park Service and the Florida Department of Transportation are retrofitting the Tamiami Trail/US-41 highway that runs east-west through the Everglades connecting Miami to Tampa—"a dike through the heart of the Everglades since the 1920s"[67]—to allow water through by installing culverts in some places and replacing roadbed with the new Everglades Skyway bridge system. Other projects include removing levees and dams and backfilling canals, buying out and converting corporate agricultural land to water storage, and constructing new reservoirs to conserve water for increased supply and flow within the ecosystem during dry seasons. The latter are understood to be critical to designing the system to withstand climate change-related extreme precipitation fluctuations. While environmentalists and engineers agree the ravaged ecosystem will "never function again without engineering assistance,"[68] the hope is that the eighty-five-plus projects that constitute CERP will be sufficient to correct the "quantity, quality, timing, and distribution"[69] of freshwater in the region, at least enough to revitalize the ecosystem and, in the critical infrastructure imaginary, also repel seawater.

For the Everglades to function as a critical infrastructure able to mitigate saltwater intrusion into urban drinking water supply, they must be redesigned and made to do what they are understood to do naturally: flow. Valued here is not only the drinking water the system provides but also the crisis management functions that water's natural movements and pressures will, it is hoped, fulfill. Like urban resilience visions of cities as integrated socio-ecological systems, there is a redemptive quality to this imaginary of transforming the Everglades into an infrastructural system to help administer Miami, in that it promises both to heal the destructive modern city–nature binary as well as the crises this relation is understood to have generated. As hydrologist William Nuttle, describes it, "we started in South Florida by declaring war on the ecosystem," and CERP is not "restoration" but "restitution."[70]

Characterizations of resilience as a fluid form of management that allows rather than blocks natural flows and processes—allowing and designing nature to "do its nature thing"[71]—are common. Here we might think of another paradigmatic Anthropocenic infrastructural nature: oysters. To combat sea rise and flooding, designers in New York now seek to re-create bivalve natural life cycles (layering atop one another from birth and beyond into death, creating wave-attenuating reefs) not to stop flooding along the city's coasts, as would a seawall, but instead to manage sea rise and flooding by letting water in, while attenuating and managing its impact.[72] In Braun's analysis, such natural infrastructures seem to entail "a doubling of neoliberal forms of government. If, as Foucault famously explained, liberalism posits the 'naturalness' of society and this naturalness must be 'let alone' since attempting to change it through law or regulation would only make things worse, today the 'natural relations' of nature must also be let alone."[73]

Similarly, depictions of nonhumans as bearers of critical powers rather than as inert objects are now common among academic and cultural theorists. As do many urban designers, they understand the Anthropocene as demanding new epistemologies that correct anthropocentric modern perspectives. Here we might think of attempts by Jane Bennett to disrupt the partitioning of the world into inert nonhuman matter and lively humans by lavishing attention on the "thing-powers" of nonhumans like worms or trash, which she argues constitute vibrant political actors in their own right, part of human collective publics akin to what Bruno Latour calls a "parliament of things."[74] Anna Tsing's *The Mushroom at the End of the World* also comes to mind, with its argument that Enlightenment views of nature as a passive backdrop ignore the lessons from entangled, lively natures, such as the matsutake mushroom whose "riotous colors and smells" thrive in disturbed landscapes.[75]

Rarely discussed in these readings, however, is the fact that making nature do what it "does" and making it perform in the way imagined requires extensive work by human experts and managers. If making nature into critical infrastructure is a biopolitical project, it thus also involves a kind of eco-governance or "geopower"—Timothy Luke's term for the application of biopower to the nonhuman world[76]—in which it is humans trying to make nature "be" in a particular way, not according to a social norm or as a resource but in the way they imagine is natural to nature. In the Everglades, making nature into infrastructure is not a matter of enrolling an existing nature, as "thing power" theorists and resilience discourses often imagine. Instead—somewhat counterintuitively, since the idea is to let nature be nature—it requires both actively reimagining and constructing the existence of the Everglades in specific terms amenable to infrastructural functioning. This geopolitical reimagining of the Everglades, as we have already seen, involves casting hydrological processes in terms once reserved for technical infrastructure. But constructing the Everglades as critical infrastructure also requires ordering technologies to make nature "do what it does"—that is, to be as powerful as it is imagined to be—and securing a specific version of nature's existing so that it functions and flows as ecologists and water managers hope it will.

Not let alone at all, Everglades infrastructure is in reality a large-scale experimental engineering project, one that brings to a crescendo what Ogden has described as the bureaucratization of the Everglades via technocratic water management.[77] Enrolled in this effort, nature is taken as an object of specific ways of governing particular to water, concerned just as earlier water management was, with its "particular qualities and force" and targeting its "movement through space," as anthropologist Jessica Cattelino characterizes modern Everglades governance.[78] And while Everglades freshwater, even though highly managed, is understood as "natural," the salt water it is intended to repel is in contrast seen as unnatural, an "intruding" and "invading" human-made nature. Everglades as critical urban infrastructure is thus a matter of designing and using one nature, perceived as natural (pre-drainage freshwater flows), to govern the incursion of threatening Anthropocenic unnatural natures (saltwater incursion inland propelled by sea rise and thus global warming). Here it is less a matter of "don't work against nature, but with it"—the well-known Buckminster Fuller maxim designers frequently used to describe resiliency infrastructure—but, as hydrologist Rene Price proposes, "fight water with water."[79] Whereas modern infrastructures can be built anywhere regardless of context—the infamous Robert Moses highway thrown down upon entire neighborhoods—Anthropocenic infrastructural nature is a context-specific governmental approach that enrolls the local forces of the Earth, flows, or life cycles and brings them into the city to circulate and move in such a way as to govern, administer, and control the movement and intrusion of other natures set in motion as part of the Anthropocene. By reengineering nature in such a way that it does its "natural," prehistoric, *Holocene*, thing, perceived natural processes will govern the risks *Anthropocenic* socionatures now pose to urban life.

City and Nature in the Anthropocene

In the search for new modes of governing cities in the Anthropocene, infrastructure is being redefined. Unlike the past—when nature was seen as a resource or dump, and infrastructure evidenced humanity's power to tame nature—today designers, planners, and scientists define nature as infrastructure, seeing its natural processes as a valuable risk-management solution. In valuing nature's capacities this way—a view that resonates with depictions by critical thinkers interested in "thing power"— designers and city government seek to make nature *be nature*. This revaluation is reversing long-standing perceptions of ecosystems like the Everglades' relationship to the city. Rather than being defined as the city's radical exterior—as it is still envisioned on maps of the Miami-Dade Urban Development Boundary, which starkly separates nature from densely developed city—in this emerging imaginary, the Everglades are being *brought into* Miami, operationalized as one of its many critical management systems. Like many other transvaluations that underwrite governmental and design policy in the Anthropocene—such as governments' reversing of colonial frameworks by incorporating Indigenous knowledge in climate policy[80]—this view of nature as infrastructure can easily be seen as correcting erroneous modern urban planning and its hubris. Rather than using nature in the twentieth-century sense as an object or "standing reserve," now nature is operationalized as a complex adaptive system, made of circulating flows and feedbacks, coupled with other critical infrastructural systems. Nature as infrastructure and the attendant idea of ecosystem services are celebrated as promising ecological solutions to modern nature/city binaries and new climate change risks. But Anthropocenic infrastructural nature, as I have named this approach, is neither a return to prehistoric nature nor a harmonious resolution of modernity. Constituted at a moment when cities are increasingly seen as valued objects whose very existence is threatened by climate change, Anthropocenic infrastructural nature is instead an historically specific risk-management response to the problematization of urban climate risk and outdated modern governance approaches based in human/nature binaries—a new biopolitical and geopolitical infrastructural strategy seen as a means of solving both.

Operationalized in this way, natural flows and processes are being taken hold of and brought into a broader eco-cybernetic urban governmental assemblage, the envisioned aim of which is to administer and secure urban systems. In this imaginary, the immense Everglades ecosystem becomes a kind of environmental-urban "stack"[81] and Miami becomes a city of coupled socio-ecological flows and feedbacks, not just in the sense of natural resources to supply the city, as in Matthew Gandy's and Maria Kaika's accounts of modern cities like New York or Athens—in which vital resources like water flow through pipes and out of taps, part of a largely invisible but essential technonatural urban metabolism—but an eco-cybernetic exchange of Holocene flows against Anthropocene flows, within a broader context of ongoing crisis.[82] Here, Anthropocenic infrastructural nature is a technology of urban government that parcels out useful natures (freshwater) from hostile or "pathological" natures[83] (human-induced saltwater intrusion), to mobilize or control them accordingly. The aim is to govern and administer risks posed by Anthropocene natures (seen as unnatural) in order to secure urban systems (portrayed as as natural as the natures they enroll for protection). This demarcation of threatening, unnatural Anthropocene socionatures like sea-level rise and what are perceived as presocial, natural natures produces an additional powerful effect: what is forwarded as most natural, and thus unquestioned, is the very urban system itself. Bookended by threatening Anthropocene nature on one side and salvific Holocene nature on the other, the urban is more naturalized than ever, unquestioned as a good in need of protection and object to save, as natural as the freshwater flows on which it relies and which it now seeks to secure.[84]

Ultimately the Everglades-as-infrastructure is of an especially existential nature, geared not only toward common management of this or that urban crisis or problem (crime, disorder, etc.) but also toward the possibility of saving the city at all. Infrastructure is of course biopolitical, a matter of securing key vital lifelines that keep laboring bodies "available to state and capital."[85] But without drinking water, it is not just that the labor power of individuals would be less available to state and capital, nor that some populations would be at risk of death or sickness. What is at stake in saving the aquifer is saving the *city* itself: without drinking water, scientists and media forewarn, one need not wait until the seas swallow Miami's famous art deco hotels and condominiums to see the death of the city. This risk is compounded by the fact that water use and demand in the area is increasing.[86]

Still, like other failed infrastructure projects, such as Paris's never-built automated train system, the Everglades simply might not function in their imagined governmental role.[87] Unlike building a road or highway, making nature be infrastructure occurs within the social, ecological, and technological legacies of the Anthropocene, but it also must compete against the epoch's present and future trajectories of climate change and environmental destruction. This brings us to a different interpretation of critical theorists' point that nonhumans are active, autonomous agents, not raw material for human needs. Their "agency" includes refusing to work or be enrolled in governance, by existing in ways that resist functioning as infrastructure. But acknowledging the autonomy of nonhumans also means recognizing the equally likely possibility of their not being powerful, by not existing at all. If the Everglades do not function, it might not be only because the ecosystem's existence has no regard for governmental purposes, but because the Anthropocene's onslaught of social and environmental destruction has rendered its existence impossible to sustain. CERP's restorative projects were designed based on precipitation and climate observations made between 1965 to 1990 and the assumption that past climates "captured the full range of variability expected throughout the 21st century."[88] But today the ecosystem is at what scientists describe as the "forefront of transition," with scientists in agreement that sea rise is accelerating and that CERP is in "a race against rising seas."[89] As a National Academy of Sciences panel report to Congress on restoration efforts concluded, "It is clear that the Greater Everglades of 2050 and beyond will be much different from what was envisioned at the time of the CERP conceptual plan."[90] Despite restoration efforts,[91] saltwater intrusion is continuing, acting as a "silent killer"[92] causing mangroves to retreat inland in a "death march" away from the sea;[93] reducing plant cover along the southeast saline coasts to a sparse, low productivity zone that appears white in aerial photos;[94] pushing pink-plumed roseate spoonbill, long seen as an indicator species of Everglades health, to relocate their nests; and causing the ecosystem's peat soil—the carbon-sequestering backbone of the system, accumulated over thousands of years—to break down and wash away.[95] Many scientists who have long devoted their research to Everglades revitalization argue that the ecosystem is already at a tipping point and that stopping saltwater encroachment and eventual submergence of the ecosystem is unlikely in the long term.[96] Water managers envision restored Everglades flows pushing back against sea rise, providing enough pressure and weight to keep the invading sea at bay and the city resilient. But if peat soil collapses and ground surface is lost, "then that's it," hydrologist Rene Price concludes. "You have the transgression in, the coastline disappears, and you have the collapse of the systems behind that line, and over time it becomes impossible to maintain a marsh system of any kind."[97] In that scenario, the ecosystem will be overwhelmed by the sea, its soils, mangroves, alligators, replaced by an open saltwater expanse.[98] In this scenario, ecologists suspect that the Everglades may shift from a carbon bank to being a carbon emitter, its peat releasing carbon dioxide into the atmosphere, not only not functioning to increase resilience but instead further propelling global climate change.

No amount of governance can make nature be. And ultimately for many reasons the imagined functioning of the Everglades as infrastructure may not be possible. This is perhaps especially true in the Anthropocene, where governance takes on an additional dystopian dimension in that the natural powers on which urban security now relies may be undermined by the effects of the very industrial liberal systems and relations such governance techniques seek to secure.

4
BEYOND RESILIENCE
ISLANDS, URBICIDE, AND INFRASTRUCTURAL DELINKING

> We are already entering what James Lovelock terms "The Next World." Cities, the new human homelands, will carry our species through this transition.
> —Brendan Gleeson, *The Urban Condition*

Urban Resilience Forever

From dramatic aerial shots of the tsunamis that triggered the Fukushima nuclear accident and the threatened coastline of southern Louisiana to slow-motion tours through the urban poverty of Medellín and Detroit, the Rockefeller Foundation's film *The Resilience Age* invokes the precarity and threatened nature of urban life in the twenty-first century, ominously declaring this "the resilience age" and that "there is no other way."[1]

Urban resilience is currently the most widely accepted approach to coastal adaptation in the Anthropocene, one embraced ubiquitously by governments, planners, and designers. This imaginary of cities, once equipped with adaptive infrastructure, able to perpetually absorb and govern extreme events and threshold crossings, thus allowing urban spaces to maintain their basic structures and functions, is widespread. Assumed in the resilient cities vision is that the inexorable development of urbanization—at least as we know it—and the Anthropocene will remain necessarily companion processes. In the preceding chapters we have considered multiple infrastructural and imaginal experiments underway in Miami to secure the city's resilient future: the $650 million resiliency infrastructure program, "Miami Beach Rising Above," with its new industrial pumps, elevated streets, and seawalls; restoration of Everglades freshwater flows to push back saltwater intrusion into the urban drinking water supply; and, at the community scale, urban futures scenario exercises designed to increase the capacities of the city's poor and vulnerable to prepare for and recover from extreme events. We also saw how these projects are transforming and, in some cases, redefining urban space and life: here elevated neighborhoods are the gray infrastructural complement to emergent Anthropocenic forms of infrastructure, a once-technological category now expanded to include nature and humans. While designed within the city's specific spatio-temporal-economic-geological context, Miami's resiliency strategies exemplify the broader resilient cities paradigm that now dominates urban planning and design.

As envisioned by proponents and critics alike, as the "deus ex machina" of the Anthropocene, as geographer Kate Derickson puts it,[2] the resilient city is a city that, by both allowing and administering ecological crises, can still be saved. Imagined across urban resilience initiatives is a future city that, although inundated by seas and storms, remains both viable and governable. In Miami, water will come—in the streets and in media imaginaries of the city—but through gray, green, and social infrastructures added incrementally over time,[3] its impact can be anticipated and managed. Existing urban infrastructure and terrain will be recalibrated, retrofitted, or supplemented, but broader system parameters will be maintained amid turbulence. Circulation of capital, tourism, and investment will be secured by buffering the impacts these structures create and are tied to. Through a bricolage of modular, integrated infrastructures able to absorb and even gain from extreme events, the global "city on the edge" will ward off its projected end and even thrive while seas rise up through its foundations and waves crash against its shores.[4]

The urban resilience imaginary has footholds and influence far beyond Miami, of course, and in nearly all domains: as we have seen, it is shaping contemporary visions of what nature is, what it can do, and its relationship to urbanization. Importantly, the resilience imaginary is also recalibrating how human capacities are understood. In the resilient imaginary, large-scale ecological and technical infrastructures work together with situated local practices and know-how of the poor and vulnerable—now termed critical "social infrastructure"—together constituting an eco-cybernetic, adaptive urban network. This revised spatial-governmental form is seen as promising a future city immunized from extreme press and pulse events, both ecological and social. This is a new urban security apparatus that reproblematizes, transforms, repurposes, and redirects past technologies of governance, while supplementing, not replacing, them with heterogeneous new practices of anticipating and attenuating the risks industrial liberal regimes produce. Rather than twentieth-century visions of order, the resilient city will be a "crisis city," perpetually vulnerable to compound risks (because resilient infrastructures do not replace but supplement existing industrial infrastructures).[5] But it is still a governable, savable city. With crisis as a constitutive backdrop, resilient infrastructures enable cities like Miami to remain financially and physically viable, subtended by and interlinked with the critical flows of capital, people, and commodities that constitute liberal industrial global urbanization. In resilient urbanism, what is seen as still possible is the continued reproduction of the city's basic social, economic, and geographical parameters amid ecological crisis. This future vision is aptly summed up in the name of Miami's $400 million resiliency infrastructure bond: "Miami Forever."

After Resilience: Extreme Adaptation

Whether loved by practitioners or hated by critics, resilience is often portrayed as the horizon and ne plus ultra of urban life in the Anthropocene, beyond which little else is discussed or envisioned. As Hillary Angelo and David Wachsmuth note, "climate change has (thus far) offered no obvious 'less city' solution."[6] But Anthropocene crises have begun to undermine this imaginary. Meditating on the challenges of the twenty-first century, Roger Keil asks:

> As we scramble to stay on top of yet the next record hurricane or typhoon hurling themselves against coastal mangroves or misplaced developments on shore, we have arrived in an era of permanent catastrophe. . . . In this situation, as once again even the most sophisticated flood control systems fail . . . we ask ourselves whether cities . . . are up to the task that lies ahead?[7]

For some planners, scientists, and water managers, the answer is no. For them, current resiliency efforts will be insufficient to manage the region's projected sea-level rise.[8] To reiterate, that is 10–17 inches by 2040; 21–54 inches by 2070; and 40–136 inches by 2120.[9] Like many resilience projects that orient only to a mid-twenty-first-century timeline, Miami Beach's resiliency infrastructures are only designed for the next several decades. After that, in geologist Harold Wanless's frequently cited view, the city is "doomed."[10] Geochemist Henry Briceño agrees: Miami, he predicts, is "the future Atlantis."[11] Such phrases and predictions are repeated often. Influential water- and flood-management expert Henk Ovink, for example, has also labeled Miami "the new Atlantis" and proclaimed that "Miami will no longer be a land city, but a city in the sea."[12]

When faced with such prognostication, echoed frequently in media and art produced in and about the city (see chapter 1), for some Miami's doom appears inevitable. As summed up by former Speaker of the Florida House of Representatives and former program director for innovative programs at the Institute of Environment at Florida International University Tom Gustafson:

> Given current South Florida elevations, GHGs already in the atmosphere, the likely acceleration of such GHG emissions

for the next twenty years and more, and expected sea level rise feedback mechanisms, there is little likelihood that South Florida, as we know it today, [will be] habitable by the year 2100 AD.[13]

Gustafson, a Greta Thunberg supporter and impassioned climate action advocate, has responded to Florida's refusal to act on climate change, suggesting—during the tenure of climate-change-denying former governor Rick Scott—"I have a solution for that. We need to all march up to the capital in Tallahassee and burn the fucker down. That's the only way we're gonna save South Florida."[14] Like many designers engaging Miami,[15] Gustafson argues that more dramatic transformations will be necessary to maintain conditions for human habitation in the region over coming decades. In academic conferences, design proposals, planning workshops, and informal conversations, he and several other practitioners in one way or another foresee the necessity of bulldozing old buildings and physically transforming the city into an infrastructurally linked, self-sufficient zone of high grounds, connected like the Florida Keys by new elevated bridges.[16]

As envisioned by Gustafson, building the "Islands of South Florida" will require converting nearly all of Miami's currently developed land into fill, which will be used to create a series of new elevated islands. On the islands will be high-rise buildings, with space allocated for luxury as well as a service workforce. The rest of the South Florida population will migrate out via managed retreat. The artificial islands, in Gustafson's vision, will be based in localized food and materials production, renewable energy, and water-based lifestyles. "What lies ahead," Gustafson emphasizes, "is limited only by our will to survive, commitment to long term planning, desire to establish a just and egalitarian society, and efforts to continuously expand the horizon of human endeavor."[17]

South Florida's Island Imaginaries

The Islands of South Florida vision (hereafter abbreviated as the Islands) is only a vision, and indeed one that does not currently have extensive purchase in or beyond Miami. It is not being constructed and it is not being funded. Nevertheless, during my research in Miami this proposal for South Florida's future stood out to me more than any other, suggesting as it does a different planning and design approach and future mode of life than those thinkable within standard resilience imaginaries. Both bold and extreme, integral to the Islands is a vision of dismantling existing urban geographies and radically transforming how humans live in South Florida, how they manage and design their built environment—including the very ground itself—and how they relate to the natures that define the region as well as to global infrastructural networks. In opening the question of transformation within all of these dimensions as a matter of *islandizing*, the Islands proposal also speaks to a range of adjacent visions frequently discussed by experts in and about the region—after midcentury, there will only be islands in South Florida—and resonates with other proposals that take a similar islandizing direction.

Consider, for example, the autonomous secession of South Florida from the rest of the United States envisioned as one possible future for the flooded region by art, architecture, and urban design collective Alliance of the Southern Triangle (A.S.T., composed of Diann Bauer, Felice Grodin, Patricia Margarita Hernández, and Elite Kedan). In this vision, secession is portrayed as a process of collaborative relocalization via technical experimentation. The group's 2019 film *Prologue: Politics as Palliative Care of the Species*, assembled by artist Diann Bauer, features a press statement inviting partnerships with oceanographers, artists, planners, engineers, bankers, and other practitioners to develop experimental techniques for terraforming, water purification, plumbing, wildlife corridor creation, land reclamation, and vertical agriculture necessary to enact South Florida's breakaway as seas rise.[18] In this vision, the idea of assembling alliances based on adaptation techniques and the building of human capacities and potential are understood as joined. As A.S.T. puts it in their essay "Protocols for the Phase Transition,"

> Adaptation is not "survival of the fittest"; it cannot be derived from market values of individuated neoliberalism. Adaptation is a dynamic force for coalition building; it is diverse capacities coalescing into a distributed fortitude that operates across a system. We . . . become the adaptation we need to be, that we need each other to be, by bringing together our diverse capacities. It is within our synthetic capacities that our greatest potential emerges—socially, technically . . . alliances . . . can flip crises into essential transformation.[19]

We might also think of architect Gabriele Filippi's "Green Water Village," the winner of Arch Out Loud's 2022 Miami Floating House competition, which invited proposals for futuristic floating housing solutions to address the city's housing affordability crisis as well as the threat of sea-level rise and flooding (Figure 14).[20] Situated in Biscayne Bay adjacent to the rapidly gentrifying Little River neighborhood, Green Water Village imagines a floating network settlement of hexagonal shaped artificial islands suspended on empty caissons with homes, gardens, and gathering spaces connected by wooden passageways. Infused by biophilic design approaches, the floating islands are arranged in an organic pattern, each dominated by soaring glass and metal cage-like structures covered in greenery and topped with wind turbines and masts, with prefabricated living structures built out of wood, glass, metal nestled beneath. As in the Islands of South Florida, Green Water Village utilizes renewable energy and rainwater and humidity capture. Explains the proposal, "Miami, future. The unstoppable climate change has caused the seas to rise: large portions of coastal cities are about to be submerged and there is a need to build new floating villages: new homes, new routes, and new ways of living and surviving. New places

Figure 14. Gabriele Filippi Architetto's "Green Water Village," the winner of Arch Out Loud's 2022 Miami Floating House competition, which invited proposals for floating housing solutions to address the city's housing affordability crisis as well as the threat of sea-level rise and flooding. Copyright: Gabriele Filippi Architetto, Original Publisher_ Arch out Loud. Date of first copyrighted publication: December 30, 2022.

of aggregation are created in continuity with suffering cities: we must resist!"[21]

Finally, islandization is also a key pillar of "Salty Urbanism," an aqua-urban design toolkit spearheaded by Brooks + Scarpa principal architect Jeff Huber for adapting coastal communities in South Florida to increased salinity and "an inevitable future of saturated landscapes."[22] Funded by a NOAA Sea Grant, Salty Urbanism is part of ADaPT: Adaptation Design and Planning Tool for Urban Areas in the Coastal Zone, a collaboration between Brooks + Scarpa and interdisciplinary teams from Florida Atlantic University, the University of Kansas, and the University of Southern California that utilizes the neighborhood of North Beach Village in Fort Lauderdale (located within the Miami metropolitan area, north of Miami-Dade County) as a test study. A guide for extreme future adaptation of the built urban environment to sea-level rise, the project centers on coupling ecological processes and infrastructure, and the need "to rethink and retool their relationship with water—one where designing for exceedance, or rather flooding, is the norm and a new mantra for South Florida living will occur: how do we live with, over, and on water?"[23] Via design studios, the team proposed three future adaptation scenarios for North Beach Village at multiple smaller scales, each increasingly extreme in terms of its intervention into existing urban space.[24] The first, Soft Defense (The Green Jacket), features green and blue alterations to the existing urban environment: a suite of living shorelines and breakwaters, blue and green streets featuring high-tide gardens, salt-tolerant plants, pervious pavements, bioswales, and rain gardens. Pushing slightly further beyond this resiliency vision, the second scenario, Strategic Retreat, focuses on "gradual removal of urban development through relocation to higher ground on the coastal ridge," densifying urban development on higher ground, and rewilding low-lying beach and coastal areas. Finally, the last and most extreme scenario, Land Adjust (Islands and Atolls), proposes amphibious architectures, elevated and submerged living units, and islandized zones both in terms of ground and infrastructure (Figure 15).

Like the above envisioned islands, Land Adjust imagines decentralization and localized production of food, water, and energy and waste management as essential components integrated into an islandized living fabric and built environment. Equally integral to Brooks + Scarpa's dramatic rehauling of the built environment are amphibious architectures and stilted neighborhoods, water taxies and ferries, and bioremediation islands, as well as legal tools such as land adjustment through which communities can redefine and redesign their land, buildings, and infrastructure. In the firm's related but broader proposal for "Unplanning Miami,"[25] by 2050 the region is imagined as resembling 1850, with smaller communities resdesigned around the carrying capacity of rainwater, smaller footprint, localized food production, and waste management.[26] For those who do

144 | BEYOND RESILIENCE

Figure 15. Land Adjust (Islands and Atolls) scenario from design proposal for Fort Lauderdale's North Beach Village neighborhood. From Jeffrey E. Huber, Keith Van de Riet, John Sandell, and Lawrence Scarpa, "Salty Urbanism: Towards an Adaptive Coastal Design Framework to Address Sea Level Rise," *The Plan Journal* 2, no. 2 (2017): 389. Courtesy of Jeff Huber.

stay, Huber suggests, profoundly transformative architectures, technologies, and modes of thinking and living will be required:

> Nascent economies that emerge from climate change and SLR will form as the only viable response to living in a saturated and salt-laden landscape. . . . To stay requires conventional practices of land ownership and management that promote separation from ecological networks to be questioned and translated into an unplanning framework. Development of amphibious strategies through new languages of wetness, buoyancy, tethering, anchorage, hosting (scaffolding), raising and suspension are considered. Left behind are preconceived notions of environmental control, mitigation, resistance, and

permanence for a symbiotic existence, adapting to a dynamic, transitional and fluid environment.[27]

Urban Dismantling

These proposals—which I will henceforth refer to as "Anthropocene island imaginaries"—suggest glimpses of a new paradigm of twenty-first-century coastal adaptation that, while related to urban resilience, must be distinguished from it. Resilience remodulates existing urban geographies and built environments to secure present social, economic, and geographical conditions. While urban resilience departs in significant ways from modern engineering-based planning, imagining instead an interconnected bricolage of situated social, technical, and ecological infrastructures, it is still fundamentally modern because it maintains hope that the city can (and should) be saved.[28] Through an interlaced, layered suite of soft and hard infrastructures to field and absorb disaster events—mangrove buffers, modestly elevated roads, pumps, seawalls, prepared communities—existing urban geographies remain viable and governable.

Anthropocene island imaginaries, however, suggest the impossibility or at least hard limits of this resiliency vision. Across the varied proposals that constitute this imaginary, existing urban formations are understood as already doomed, destined to be swallowed by the century's rising seas. Challenging hegemonic assertions that, as the Rockefeller Foundation puts it, there is "no other way" but to become resilient, and the idea, central to resilience, that systems can endlessly endure and absorb industrial liberal society's disasters, adapting and surviving while perpetually remaining vulnerable, the islands imaginary foresees the impossibility of making the existing urban agglomeration resilient.[29] Accordingly, rather than retrofitted, in proposals such as the Islands of South Florida and Unplanning Miami it must be preemptively dismantled. Envisioned here is not elevating the city—as engineer Bruce Mowry puts it, "defend[ing] against the water at the present interface of the ocean to the land"[30]—raising roads a few feet and eventually a whole city as in Miami Beach's efforts. Instead of making existing spaces and systems resilient, the accent, at least initially, is increasingly placed on their dismantling.

The Islands of South Florida, for example, envisions the necessity of pulverizing the current city in order to use its substrate as fill for a new geography.[31] In this vision, Miami's physical ground appears as a pragmatic and existential problem, and solution: source of urban vulnerability to sea rise and raw material for new territory. This physical ground was itself recently engineered, and Miami itself is a product of extensive terraforming. Much of the city was built—largely by Black laborers from the Caribbean and American South and poor white workers—on drained swampland. Likewise, the current topography of Miami Beach, a thin barrier island in the late nineteenth century, was created in the early twentieth century as part of a real estate scheme by entrepreneur Carl Fisher. Under Fisher, workers, elephants, and mules cleared the island's dense mangrove swamps and filled in the land.[32] Millions of cubic yards of sand dredged from adjacent Biscayne Bay and Everglades soil were added atop existing limestone foundation and arranged into level terrain three to five feet above sea level.[33] Miami Beach's grid of streets, pavements, and condos was overlaid on this new foundation, built with rock brought in by barge from inland mines, providing terrain for Fisher's new industrial money crowd.[34] In fact, many of Miami Beach's most exclusive residential islands, like Star, Palm, and Hibiscus Islands—home to billionaires and celebrities including Shaquille O'Neal and Jennifer Lopez—are artificial islands built using dredged Biscayne Bay sand in the 1920s. And, not limited to the twentieth century, ground shaping also has a much longer history in the South Florida region. Evidence suggests that prehistoric inhabitants terraformed the tree islands found throughout the Everglades, elevating them through the addition of bones, shells, and sand.[35] Chokoloskee Island, located on the western edge of the Everglades, is a 150-acre, twenty feet tall in some parts, human-made island/shell mound—the largest shell mound in the country—thought to have been created by the Calusa.[36]

Carrying the bricolage elements of this approach which has long subtended South Florida forward into the Anthropocene, in Gustafson's vision, government must obtain all of the city's developed land to use as fill.[37] This would be done via multiple methods. Some land must be condemned. A smaller percentage, he suggests, can be sold to private developers and subsequently bought out. As in the limestone mining operations across Florida, conceivably this would involve clearing earthen and built materials that would not be used. Iconic palm trees would be replanted or mulched.

In this dramatic vision, the city's ubiquitous excavators and bulldozers are redirected. No longer digging ground for the expanding forest of luxury condos and hotels that crowd Miami's waterfront, instead they are demolishing buildings and waterfront landscaping both. Demolition is understood, however, as key to *salvation*—to "save our South Florida," as Gustafson has put it. Once the land has been cleared, excavators would conceivably remove soft sandy top layers. Explosive dynamite would break up hard areas of limestone. Both would be crushed into a uniform size suitable for use as fill. The land itself would be used as fill to create a series of massive, elevated islands. Atop these human-made islands will stand high-rise buildings, inhabited by residents who enjoy water-based lifestyles and with space allocated for luxury as well for a service-sector workforce. The rest of the South Florida population that does not stay will be relocated through managed retreat.

Similarly, albeit from a more concrete design angle, to facilitate the dramatic reformattings required to remain in South Florida with six feet of sea-level rise, Brooks + Scarpa's Huber has proposed the creation of a Department of Unplanning, tasked with planning ahead and designing for retreat and evacuation of soon-to-be-perpetually-flooded areas, through the considered dismantling and decommissioning of existing infrastructures, analysis of the plants and land typologies that can remediate the contaminated lands, returning of lands to nature, and transitioning institutions such as water, energy, public schools, and police to a future of smaller populations and rewilded nature. "Unplanning Miami," as the project is named,

> repositions how we live, adapt and transition urban development, especially when we must leave or adapt to the land that is no longer high and dry. Therefore, just as governmental agencies have planning departments, so too, should they consider the establishment of a Department of Unplanning. Ultimately, decisions will need to be made on which neighborhoods are abandoned and which will be modified and adapted. Rising groundwater tables compound this risk due to the region's porous limestone geology that will flood far from coastal shorelines and, ultimately, cause a complete loss of the region's drinking water supply. Policy and design solutions are not truly considering the necessary transformation required to live, work, and play within a saturated coastal environment. The old paradigm of flood management and control will need to change from prevention to acceptance. Abandoned development will need to be prepped and retrofitted with a living infrastructure capable of cleaning the environment as populations shift. Furthermore, population decline will occur as businesses and individuals decide the costs are too high to maintain regional assets and decommissioning (unplanning) of the built environment will come into more clarity as a pressing challenge. These challenges will transform regional tourism, housing, waste management, energy and food production infrastructure.[38]

In such proposals, the urbanized region, so recently built in the Anthropocene's ascendant phase, comes to appear as a failed experiment, to be dismantled at the epoch's decline.

Anthropocenic Urbicide

How might critical urban thinkers understand these imagined productions of new archipelago territories? Certainly, one might consider island imaginaries, especially the attendant vision of dismantling urban spaces, via the planetary urbanization lens.[39] Though less considered within planetary urbanization literature, urban climate adaptation is already generating its own novel urban sites and processes in Miami and elsewhere. Like Miami Beach's elevated roadways or the Everglades-as-infrastructure, island imaginaries could also be read as visions of novel territorial formations and rescalings generated by the endless dialectic of planetary urbanization and creative destruction. One might ask, given its hurricane exposure, oppressive heat, and, once bulldozed, lack of attractive beaches, why rebuild in this location? The Miami answer might be, why not? South Florida's current environment was built as real estate speculation upon wetlands deemed uninhabitable by governments and planners. Miami Beach's beaches are themselves artificial. As we have seen, the city is grounded on spectacular imaginaries of subtropical life that attach people to the region and that, retooled, could easily fuel another round of libidinal-financial investment.

Following such a reading, future imaginaries such as the Islands or Unplanning Miami might appear to critics as a form of anticipatory creative destruction or—given the accent placed on dismantling—urbicide, albeit understood in the new context of the Anthropocene, with its emerging imperative of climate adaptation (although, to preview my argument, I will suggest the need to question this reading). Traditionally understood as the military tactic of demolishing ("killing") cities, and frequently associated with counterinsurgency, urbicide has long been a strategy for attacking an insurgent force's "urban foundations"[40] and has been used to describe the American military's walling off towns and urban districts in Iraq or the Israeli Defense Forces' bulldozing of Jenin camps.[41] Urbicide is often a process of creative destruction tied to the construction of new spatial forms:

in Iraqi cities, for example, the American military sought to install a liberal regime of life and governance. But as Stephen Graham has documented, urbicide is linked not only to wartime but also to peacetime efforts by neoliberal governments, especially in the Global South, to rebrand as global cities by bulldozing slums, evicting dwellers, and building financially lucrative highways, high-rises, and stadiums.[42] Urbicide, then, has long been part of capitalist restructuring and urban development—and as Michael Roman-John Koscielniak demonstrates of the Detroit Demolition Program, such demolition processes are themselves part of broader political-economic networks[43]—and various urbicidal processes might also be seen as integral to all urban resilience projects, in that they frequently entail demolition of "outdated" infrastructure or urban space and replacement with more appropriate adaptive infrastructures and urban forms.

Along similar lines one might also think of geographer Kasia Paprocki's discussion of out-migration from Bangladesh's rural southern coast to Kolkata's peri-urban slums.[44] While the above-described Anthropocene island imaginaries are but hypothetical, Paprocki's exceptional work explores how material strategies of anticipatory dismantling are already occurring in rural coastal Bangladesh. There, portraying the region as subject to "inevitable" sea rise—based on climate *imaginaries*, she emphasizes—international financial institutions have dismantled infrastructures and lands through which rural communities had sustained themselves. Through preemptive salination of waterways, the freshwater rice farming on which these communities relied has been destroyed and forcibly replaced by commercial shrimp aquaculture, unlocking new profit sources while dispossessing communities. Further cementing dominant visions of the future twenty-first-century as a necessarily urban age, this process has forced migration from Bangladesh's rural southern coast to Kolkata's peri-urban slums and driven farmers to become labor power in industrial urban and emerging IT centers. "Anticipatory ruination," as Paprocki labels this process, is "a discursive and material process of social and ecological destruction in anticipation of real or perceived threats."[45] The concept of anticipatory ruination is especially powerful and pertinent for understanding many ecological and climate-adaptation-related urbanization processes. Think, for example, of the recent rise, alongside climate resiliency projects like those Paprocki describes, of green austerity and relocation measures being promoted in the name of climate adaptation. Here, we might also think of the Dutch government's recent seizure of farmlands in the name of climate change mitigation or the growing chorus of calls to abandon supposedly already-obsolete coastal areas such as southern Louisiana and relocate whole populations to newly designated "climate refuge cities."[46]

If read as just another instance of these processes, then, the dismantling envisioned as key to the above island imaginaries might appear as a kind of Anthropocene urbicide—albeit in a very different sense from the term's traditional usage—via dismantling an existing urban form—its infrastructure and geography—and converting urban ground into a new spatial formation: twenty-first-century aqua-urban island territories. Read thus, critics might see these island imaginaries as reinforcing visions of a future in which planetary urbanization's "Neo-Haussmanization," as Andy Merrifield describes it, an implosion–explosion process, now includes dismantling entire cities.[47] In this reading, the imagined physical dismantling of the city and transubstantiation into fill and elevated living spaces appears as a dramatic "spatial fix" to manage crises and maintain existing social and economic order. Katherine McKittrick has argued[48] that understandings of urbicide focused solely on the destruction of urban infrastructures abstract from the human impacts, and taking this perspective, critics might also extrapolate that at issue here would not just be the moving of dirt and sand and trees but also the demolition of the structures and landmarks of human life in Miami since its recent creation, along with the unremarkable, unrecorded fabric of all the lives lived in the hot, humid city "floating," as Joan Didion once put it, "between a mangrove swamp and a barrier reef."

But are such negative trajectories the only ones possible for and appropriate to Anthropocene island imaginaries and their dramatic adaptation and even urbicidal visions? Is an externally imposed "anticipatory ruination" of infrastructures and spaces through which communities in regions enframed as climate change hotspots sustain themselves, and corollary destruction of communities and local attachments, such extreme designs' necessary ethical fabric or outcome? Or, might there be other possibilities, including ones involving the imagination and activity of residents themselves?

Maybe instead of more of the same, in a world thawing and coming undone at the seams, Anthropocene island imaginaries can also be understood as signaling something else—a unique response and horizon specific to our time.

To Cut

Reading island imaginaries as (imagined) instances of planetary urbanization is important, but there are also other interpretations possible. Significantly, the futures laid out in these island imaginaries are envisioned not as a mere continuation of contemporary planetary urbanization but, in nearly all the above-described proposals, are instead articulated in attempted opposition to the historical-geographical relationality it entails. Along with pointing to the ever-evolving process of creative destruction that is urbanization itself, at core, planetary urbanization describes our contemporary global order defined by unbounded infrastructural interconnection, integration, and relationality. Emphasizing the historically specific character and drivers of global urbanization, the concept captures how contemporary capitalist processes operationalize distant, heterogeneous, seemingly non-"urban" sites to support agglomerations of urban life with energy, food, water, and other resources (as in, for example, the operationalizations of the Everglades discussed in the previous chapter).[49] Indeed, today we live in an era of global urbanization, made through the linking via roads, pipelines, logistics and supply chains, cables and distribution routes of nearly all social and ecological relations into an increasingly congealing, yet uneven, variegated, and constantly morphing, worldwide *infrastructural* fabric.[50] Thus for many urban theorists and many urban geographers too, understanding today's urban condition requires moving beyond "obsolete"[51] bounded and binarist analytics that view the urban as an "enclosed territorial arena,"[52] and toward relational approaches more appropriate to twenty-first-century realities.[53] Following from this perspective, Miami is an urban agglomeration, but one deeply interlinked with this far more expansive global infrastructural network.

Relational analytics are indeed hugely necessary for mapping contemporary neoliberal globalization's infrastructures, crises, and governmental modalities. It is for this reason that the updating of inherited urban thinking by planetary urbanization scholars has been so important and impactful. After all, infrastructural interconnection has been central to western neoliberal governance for decades.[54] After September 11, securing interconnected critical networks came to be seen as especially paramount, with American military analysts like Thomas Barnett dividing the world into a "functioning core" and a "non-integrating gap," the latter defined as "disconnected from the global economy and the rule sets that define its stability."[55] "Eradicating disconnectedness," argued Barnett in his much-discussed *The Pentagon's New Map*, "therefore becomes the defining security task of our age,"[56] with failure or refusal to integrate into the global economy and its rule sets enforceable by military action. Indeed in architect Ross Exo Adams's estimation, it is precisely this global network of flows that resilient urbanism is deployed to secure. For Adams, resilient urbanism constitutes a "sinister" paradigm in which what is actually "'sustained' is, at best, a process without beginning or end, and whose only promise is the managed expansion of the urban regime"—a single, global urban expanse without exterior amid the naturalization of environmental collapse as a shared, inevitable global condition.[57]

However, even if it defines the world today, and even if urban resilience approaches envision its eternal continuation, is interconnected planetary urbanization actually the "telos" and "final frontier" of life in the Anthropocene?[58] Amid the current interest in mapping, analyzing, and building complex, open, entangled urban-ecological systems that largely dominates urban studies, planning, and design, this is frequently portrayed, if even implicitly, as if it is the case. As David Madden remarks of globalization and the global urban, "the qualities of enclosure, finality and totalistic lack of differentiation are central to it"[59] and, connecting with philosopher Jean-Luc Nancy's work on struction and glomus, describes the latter as "not the creation of a world . . . but instead just a 'piling up,' the accumulation of junk, an unprecedented geopolitical, economic, and ecological catastrophe."[60] In this reading, global urbanization's operationalizing of the world is part and parcel of the Anthropocene, which is portrayed as precisely such a catastrophic piling up, as in the film *Anthropocene: The Human Epoch,* which tours the ruinous effects of operationalizing distant sites to serve urban life: deforestation in British Columbia, urbicidal German coal mines, metallurgical pollution in Russia, massive landfills in Kenya. As discussed in this book's introduction, in these and many other portrayals, the urbanization of the Earth is often portrayed as a necessarily companion process to the Anthropocene, both its historical and contemporary driver and outcome, as well as its present and *future* characteristic.

South Florida's island imaginaries, I argue, however, suggest a glimpse of other nascent possibilities. Instead of further infrastructural integration—in which the whole Earth is endlessly connected and depleted—island imaginaries envision not only dismantling existing urban ground but also delinking from global infrastructural networks and interdependencies. Here Graham's "switching cities off"—military targeting of critical urban infrastructures—takes on very new meaning.[61] Expanding beyond even Green Water Village's vision of floating islands run on renewable wind power and rainwater capture systems, for Gustafson, the Islands of South Florida must redesign and localize all of the infrastructures that subtend Miami life, untethering the Islands from global networks.[62] Here the island appears as a key Anthropocene site, albeit not evoking relationality—which Jonathan Pugh and David Chandler aptly show is typically how islands are portrayed in critical Anthropocene work—but detachment.[63] Among other infrastructural transformations, the Islands will "grow locally most or all our food; manufacture locally most or all our goods and building materials in daily use; and, secure the raw materials/durable goods we might need."[64] As with Salty Urbanism's vision of localized infrastructure and islands, the Islands are envisioned as a "Regional Refuge" where production is localized, something residents participate in themselves. Buffered by mangrove forests and saltwater-resistant vegetation, the envisioned island territory features extreme storm-resistant architecture to withstand the region's expected hammering rain, storm surge, and powerful winds. Power is generated through wind and solar technologies supplemented by fourth-generation nuclear. The Islands maintain an electronic "Royal Library of Alexander" and seed bank of crops suited to the region's long-term, climate-change-altered growing conditions. While relocalized in terms of production and distribution, the self-sufficient Islands are interconnected regionally by massive bridges built for smart electric vehicles and automated trucks to move supplies and people between islands.[65]

Similar processes of delinking are envisioned by both A.S.T.—as seen in their call for secession and analogous collaborative investigation of the arts and sciences needed to reinvent infrastructure in South Florida—as well as Brooks + Scarpa's toolkit for Unplanning Miami, which envisions an islandized territory for 2100 in which

> basic services and economies organized around a decentralized network tooled for salty and desert-like conditions creates novel approaches to tourism (ruin porn), agriculture, and scientific research. Water would be produced through harvesting and cleaning rainwater, as well as desalination. Waste would be collected and metabolized through phytoremediation

networks. Power would be generated with wave, wind, waste, and solar as a distributed and redundant system. Food would be grown within localized networks to service and provide sustenance farming for residents and visitors. Automated and autonomous vehicles would be placed into service to aid in clean up and detoxifying previously developed areas and waters. Abandoned structures become scaffolding for transitional and transformed ecologies.[66]

Here is a second core new definition of Anthropocene urbicide: strategically cutting key infrastructural links that defined global urbanization at the twenty-first century's dawn, in order to survive its twilight. In identifying global infrastructural connectivity as a key problem of the Anthropocene, and infrastructural network delinking as a spatial-infrastructural strategy for inhabiting it, these proposals are far from alone. The twentieth century saw the progressive building of transportation routes to move commodities from far-flung production sites to points of consumption, pipelines to move oil and gas from distant extraction sites, massive energy grids connecting regions. But in the Anthropocene, industrial liberal societies' dependence on "vital" infrastructure systems is increasingly understood as not only environmentally unsustainable but also as a vulnerability, witnessed for example in urban supply-chain interruptions and fuel shortages caused by recent hurricanes, Covid-19, or the Russia-Ukraine War.[67] Resilience and critical infrastructure protection strategies respond to these crises by attempting to secure existing networks. In contrast, islandization is already emerging as an Anthropocene strategy used by diverse actors to secure themselves from disasters, populations, or infrastructural networks deemed harmful to them by cutting links (while often selectively maintaining other networked connections).[68] Here, think lavish doomsday bunkers;[69] floating mansions and eco-districts—"interiorised," controlled "microclimates" embedded within cities;[70] or "future cities" like Eko Atlantic City, a corporation- and bank-funded multibillion-dollar walled private luxury "resilient" city development off the coast of Lagos, Nigeria, with its own autonomous power, water, security, and seawall.[71] Or, one might think also of the Silicon Valley–backed Praxis Society, whose plan to construct a new "hero futurist" city in the Mediterranean is intended to carve out an exit from Western society in breakdown and revitalize urbanism.[72] Or, in yet another different register, think of architect Vincent Callebaut's *Lilypad*, an amphibious "ecopolis" for climate refugees off the coast of Monaco.[73] Many of these are part of the broader embrace of designing new cities, which includes NEOM's The Line, California Forever, Telosa City, Akon City, and many others.

Taking strategic islandization even further into Anthropocene territory, the Seasteading Institute, founded by Milton Friedman's grandson and originally funded by Peter Thiel, is exploring how to set off a "Cambrian explosion" of autonomous artificial floating cities starting in the Pacific Ocean.[74] In the institute's eyes, seasteads are a solution to sea rise as well as what they see as the domineering overreach (taxation) of existing governments—a way to "liberate humanity from politicians."[75] As imagined by the institute and its startup company Blue Frontiers, autonomous floating cities will run on solar and wind power, grow food via aquaculture, and use desalination as well as their own cryptocurrency. Extending delinking to space, Elon Musk envisions SpaceX's Mars colonization as an escape hatch from an increasingly degraded Earth and its "rickety" civilization.[76] Indeed, Musk has suggested that the key "acid test" of SpaceX's Mars colonization viability will be its establishing self-sufficiency from Earth-based supply chains and life-support networks.[77] Likewise, Jeff Bezos describes his own space colonization investments as the only way to escape the limits of Earth's diminishing resources.[78]

In a different but related sense, islandizing was also on the rise during the Covid-19 pandemic, for example with controversial border closures around actual geographic islands like New Zealand figuring largely in debates around the ethics and efficacy of pandemic management. And, in an offensive rather than defensive register, we might also note the global de-networking of Russia in response to its invasion of Ukraine. But importantly, as already discussed extensively in chapter 2, democratic, bottom-up relocalization projects geared toward community autonomy, circular economies, and local infrastructure are just as equally growing in popularity. In many such efforts, islandizing—detaching from industrial infrastructural systems and entanglements—is understood as a means to regenerate the health, life quality, and sustainability of communities and ecosystems.

In visions for delinked islands—infrastructural, political, and economic islands, on land or water, Earth or Mars—the accent is placed on the process of detaching and delinking from historically and geographically specific infrastructural networks, rather than further interlinking. Notably, the "cide" in *urbicide* comes from the Latin caedere, which also means "to cut."[79]

How Not to Think about Islands

The potentially brutal inequalities and exclusions of "lifeboat cities," to use Brendan Gleeson's term, are obvious, and scholars like Mike Hodson and Simon Marvin have been criticizing the injustices of "urban ecological security" enclaves for over a decade—though they were usually concentrated on fortifying existing cities like London, not demolishing them.[80] But if we focus less on the envisioned spatial *form* of imagined islands such as those in Brooks + Scarpa's Salty Urbanism, A.S.T.'s secession, or Gustafson's Islands of South Florida, and instead on the delinking *process* perceived as key to it, we see that these islands' infrastructural relocalizations imagine neither immunizing current urban forms or processes,[81] nor fortifying existing cities, nor just creating new or future elite enclave cities as if on a tabula rasa,[82] but rather cutting key infrastructural links to create a new geographical form in opposition to planetary urbanization itself. This is a vision of building an "outside" to planetary urbanization as a strategic necessity for surviving—and regenerating life—in the Anthropocene. It is important to clarify what "outside" means here. Clearly, planetary urbanization—the concept or historical socioeconomic formation—has never encompassed or "mastered" all views or spaces (as many critical scholars have argued).[83] As Brenner and Keil emphasize, the concept of planetary urbanization is not meant as a claim that the whole planet is now covered in a single city. Rather, the point is that urbanization processes have today, just as Henri Lefebvre once predicted they would, "increasingly come to condition all major aspects of planetary social existence and, in turn, that the fate of human social life—indeed, that of the earth itself would subsequently hinge upon the discontinuous dynamics and uneven trajectories of urbanization."[84]

Thus, "outside" here does not refer to spaces or views that planetary urbanization as totalizing theory occludes, nor is it meant to offer "difference" as a "corrective" to "abstraction," as Hillary Angelo and Kian Goh have aptly put it.[85] Instead, approached from a perspective situated in the present spatial and temporal moment, islandization appears as a historically specific spatial strategy envisioned from within the historical formation known as planetary urbanization—which dominates human and nonhuman life today—via attempts to break from the networks and flows that constitute and maintain it. This is a specifically Anthropocenic vision arising in pragmatic response to perceived material problems (sea rise and infrastructural and geographic vulnerability).

On this note of contemporaneity, we might also note the potential resonance of visions of infrastructural self-sufficiency and human-nature reconnection that animate island imaginaries with urban ecological design proposals from the past, such as architect Paolo Soleri's Arcology, with its circular economies and sustainable human-ecological metabolisms.[86] Or one might even see resonances as well with Ebenezer Howard's Garden City, an early attempt to envision the sustainable integration of urban life and nature.[87] However, here too, as in chapter 3's discussion of past entangled urban-ecological systems approaches vis-à-vis Anthropocenic infrastructural nature, it is important to note that past designs such as Soleri's or Howard's were viewed as the evolution of urbanization toward sustainability and harmony, and were conceived in response to specific problems of their time: Soleri's miniaturization and frugality, responses to the urban sprawl of the 1960s and 1970s; Howard's garden cities, a response to urban pollution and overcrowding and isolation of the countryside at the turn of the century. While urban sprawl and pollution have not disappeared, today's islands proposals are all in their own ways conceived as a response to specifically contemporary, Anthropocenic problems, in particular the hard limits of climate change and existing urban geographies, as well as infrastructural fragility and a globally networked system deemed unsustainable and in some cases undesirable. "Large portions of coastal cities are about to be submerged and there is a need to build new floating villages: new homes, new routes, and new ways of living and surviving," as architect Filippi puts it of Green Water Village. Along with the rather obvious fact that the islands are envisioned as new floating territories—rather than land-based formations—we might also note that, rather than utopian new city visions, the islands are conceived as the product of dismantling and radically repurposing existing urban built environments, envisioned as an experimental process of figuring out what to do with the built and environmental legacies of the nineteenth and twentieth centuries—Anthropocene "inheritances," as Emmanuel Bonnet, Diego Landivar, and Alexandre Monnin, whose work is discussed in the next chapter, put it.[88] Thus, although carrying forth elements of past biophilically inclined urban planning and design visions, these islands need to be understood as a historically distinct imaginary devised in response to specific understandings of problems on the ground, and with their own unique vision of corresponding responses and living spaces.

Some scholars may hesitate at these islands' notions of urban dismantling and islandization, particularly if they read these envisioned processes through traditional critical lenses. For example, they might forefront the potential dark trajectories of islandization, pointing to the histories of top-down technocratic projects and dispossession and destruction that has characterized planned migration and urbicide.[89] Managed retreat, an increasingly accepted climate adaptation strategy, critical scholars argue, has led to both subtle and explicit forms of forced relocation, seen in the expropriation of informal settlements in cities like Manila and Medellín or raising of flood insurance premiums in New York or Miami.[90] Read as an extension of these lineages—and perhaps a "migration" of the "enclaving" spatial imaginary into Miami's unique climate change context—Anthropocene islandization and attendant reterritorializations could suggest a glimpse of potential recalibrations to current governmental formations that may emerge as liberal regimes seek to maintain themselves amid the century's shocks and stresses (including climate migration, which Hurricane Maria has already set in motion from Puerto Rico to South Florida).[91] Far from utopian floating cities, this might bring to mind dystopian visions of a global regime of climate austerity and exclusion.[92] On one side, as sociologist and criminologist Nigel South puts it, the "future architecture of the elite . . . promis[ing] elysian sanctuaries . . . Disneyesque escapes from the realities of the state of the planet,"[93] the world's richest 1 percent fortified against undesirable environmental and human surrounds; on the other, the rest of the world subject to ecological breakdown, dwindling resources, and migration.[94]

One might note that Miami is already home to certain possible previews of this trajectory, in the form of pandemic-era islandization of actual islands. Fisher Island, for example, is a 216-acre artificial, private, members-only island enclave located off the southern tip of Miami Beach. It is accessible only by private helicopter or boat, containing its own health-care clinic, and covered in luxury dwellings owned by billionaire celebrities such as Oprah. In 2015 the island—christened after Carl Fisher, who once owned it—was named the richest zip code in America with an average income of $2.5 million and a $250,000 initial membership fee as well as $22,256 annual dues.[95] In the first month of the pandemic, when the majority of Miamians were unable to obtain Covid-19 testing, the island, which runs its own health clinic for residents, provided rapid tests for every resident and worker on the island.[96]

If seen as yet another instance of the island-qua-enclave, South Florida's island imaginaries might then be dismissed as yet more examples of what Mike Davis and Daniel Bertrand Monk called "evil paradises," like Dubai, which for Davis was the epitome of hubris, with its "The World" gilded archipelago of artificial islands and apocalyptic luxury.[97] Or, scholars might also find much to criticize at a conceptual or even ontological level in the conceptual emphasis on delinking that infuses such islandization visions. From the relational entanglement perspective frequently used by critical theorists—an "earthbounded," eco-cybernetic view that dominates much critical Anthropocene thought and is often tied to the critique of top-down planner sight discussed in chapter 2—islandization strategies would likely be labeled backward, in the same way islands were themselves once viewed negatively vis-à-vis modern thought.[98] More specifically, islandization might be viewed as regressive because it represents a continuation of the modern desire of an agential subject to gain control over their environment, to—as in the God's-eye-view state sight from chapter 2—detach from entanglements, to insist, contra much critical Anthropocene theoretical discourse, such as that of Timothy Morton, that there is in fact an "away" from loops and disastrous entanglements.[99] Understood thus, islandization would likely be dismissed as the attempt of what Latour describes as modern humans clinging to outdated ideas of control, agency, and hubris.[100]

However, these readings can be reversed. In fact, as I will argue further in the next chapter, reclaiming urban imagination in the Anthropocene *requires* doing so. Instead of using easy, inherited modes of interpretation to understand the fundamentally new era that we are in, urban thinkers and practitioners must challenge ourselves to imagine possibilities for extreme climate adaptation that outstrip what we have known and come to expect thus far, including via well-worn critical lenses. As Shumon

Basar puts it in an essay aptly subtitled "How Not to Think about Dubai,"

> When critics bring their conclusions, a priori, to a place, it prevents what could be a productive confrontation between critic and site, in turn preventing any unexpected or unforeseeable critical responses. Predestination, instead, governs. There's no self-criticism, no threat of mea culpa, and no promise of unsettling Alterity. . . . So when do critics fail? When they form intractable judgments before they've even begun to open their eyes.[101]

After all, there is nothing necessarily regressive about delinking. As theorist Eva Haifa Giraud argues, delinking or disentanglement can be a liberatory process of subjects detaching from structures and situations that strangle them, to reweave others according to other priorities.[102] And as discussed in chapter 2, less overdetermined efforts by poor and working-class people to build local infrastructures and autonomy also abound. Notably in this vein, the Islands of South Florida proposal draws on Gustafson's ongoing work with EcoTech Visions Foundation, which, as discussed in chapter 2, is a Miami-based "disruption," "ecopreneurship" nonprofit focused on democratic, free dissemination of green manufacturing and tech skills to marginalized communities via maker and incubator spaces.[103] Considering that urbanization's interlinking of commodity and infrastructural flows into an uneven yet continuous web—the "metropolis," which the Invisible Committee describe as "a flow of beings and things, a current that runs through fiber-optic networks, high-speed train lines, and video surveillance cameras, making sure that this world keeps running straight to its ruins"—is also a technology of proletarianization and, as Adams shows powerfully, spatial *control via circulation and connection*, an islandized territory and population delinked from global networks could just as equally be democratic or emancipatory, a utopian attempt to escape manufactured vulnerability and build community autonomy via local production, sustainable energy, and water-based lifeways.[104]

Furthermore, we are perhaps accustomed to urbicide and anticipatory ruination being processes enacted by nefarious and powerful actors upon helpless populations. But suggested by Anthropocene island imaginaries like Unplanning Miami, the Islands of South Florida, or A.S.T.'s secession is a very different definition of both, in which anticipatory dismantling might come to be seen as needed and carried out by urban communities themselves. From this vantage point, instead of an endless repetition of urban resilience, South Florida's island imaginaries suggest that there could be other possibilities for human-environment cohabitation in the region, including ones imagined and enacted together by designers, planners, engineers, scientists, artists, and residents themselves. In this way, South Florida's Anthropocene island imaginaries allow us to imagine what resilience approaches rarely do: the possibility of extreme and profound urban transformation, via the audacious terraforming of wholly new aqua-urban living spaces, by and for the people living in them. The hubristic agency needed to drive such a project is not, as Anthropocene critical thinkers often suggest, an outdated relic of the Holocene. Rather than championing the mere continuation of existing systems—or worse, diminished life, the end of abundance, and visions of human life reduced to survival and infrastructure—this is a time for exploring what comes next for human life on Earth. In their own ways, each island imaginary for South Florida's future does this.

To make such imaginaries real, what A.S.T.'s "Protocols," Brooks + Scarpa's Salty Urbanism and Unplanning Miami, and Gustafson's Islands of South Florida all point toward, in one way or another, is the need for a massive interdisciplinary process of experimental research and development—a science focused on identifying the diverse array of "cosmotechnics," to use Yuk Hui's term,[105] necessary for the hubristic breaking away from existing globalized urban systems in the Anthropocene, and appropriate to living with water in the South Florida region specifically. Such an investigation encompasses all aspects of human means of existence: water, food, energy, architecture, culture, environmental regeneration, land reclamation, landscaping, terraforming, human-environment relations, and, in the Islands of South Florida and Unplanning Miami, even forms of urbicide, envisioned not in its traditional negative sense but as undoing infrastructures, built environments, and forms of habitation that no longer suit, a necessary salvo for this transformative endeavor. Far from small-scale, primitive visions of relocalization, these islandization visions evoke equally the need for massive formations as well as advanced technologies and design approaches—3D printing and automated cars as much as mangrove forests and rooftop farms productive of locally appropriate, nontoxic crops; centralized as well as decentralized planning.

Viewed thus, the Anthropocene islands imaginary is less a specific design and more an image for thought and practice: inviting us to push beyond resilience imaginaries and to seize the transformative potential of the twenty-first century, to explore the as-yet-unknown possibilities for what urban thought, design, and life can become in it. Despite what urban resilience imaginaries often implicitly assume, as Nietzsche's Zarathustra puts it, "there are a thousand paths that have never yet been trodden, a thousand forms of health and hidden islands of life. The human and human earth are still unexhausted and undiscovered."[106]

Horizons

In the end, whether positive or negative, emancipatory or regressive, islands qua sites and imaginaries of delinking are key Anthropocene spaces in need of further critical analysis. To conceptualize them it may be useful to draw on recent geographical work on the concept of territory. Building on the work of geographers Stuart Elden, Phil Steinberg, and Kimberley Peters, island territory created by delinking can be thought of less in terms of a two-dimensional bounded area and instead as heterogeneous, emergent assemblages on land or sea—"volumina," to use Peter Sloterdijk's term, made of their own technologies of living, ways of moving, geopowers, and interactions with humans and nonhumans.[107] Delinked island territories may well be regressive, xenophobic, or a strategic new means of capital accumulation. Many existing islandization efforts certainly tend this direction. But delinked island territories could just as well be something else entirely. Although it is an antagonistic concept, "territory," philosopher Elizabeth Grosz writes, is also "artistic, the consequence of love not war, of seduction not defense."[108] In this vein, is it possible to imagine emancipatory trajectories of delinking (or deliberate Galápagosization) taking shape at a comparative scale and depth of power to those of the planet's ruling classes? Will the epoch be marked by a widespread movement of peoples delinking from toxic, dehumanizing structures to create other, rich, unbounded territories, ones infrastructurally and subjectively capable of deciding how to live on their own terms?

Ultimately, the socio-ecological relationship between islandization and the broader urbanized, neoliberal planet remains to be explored and the technopolitical questions South Florida's islands imaginaries raise—would the Islands of South Florida be another Eko Atlantic City? A delinked and democratic Anthropocene refuge by/for the region's diverse peoples? Or something else entirely? Good or bad, would its attempted outside even be possible?—cannot be answered hypothetically.

What we can, however, say about the Anthropocene islands imaginary is that its delinking emphasis suggests a key point for urban scholars, which is that planetary urbanization work's foundational injunction—Brenner's call to relentlessly question inherited spatial concepts and analyze mutations in urban form/process produced by contemporary global capitalism—has not been taken far enough within urban thought and practice. Planetary urbanization and relational entanglement analytics may describe existing formations, but forecasts—often embedded in urban resilience's imaginaries—of their continuing infinitely (whether linearly or dialectically via implosion/explosion) toward ever greater enmeshment may be premature. Similarly, rather than the apex of urban climate adaptation, urban resilience may instead turn out to have been but one phase—one apparatus or paradigm of urbanization—alongside or to be supplanted by other, more extreme or transformative approaches.

After all, as Brooks + Scarpa's Huber points out, citing a Brookings Institution statistic, "50% of the built environment projected to exist by 2050 currently does not exist."[109] As climate change progresses and urban adaptive responses mutate in kind, perhaps the unique late-twentieth- and early-twenty-first-century formation known as planetary urbanization—itself superseding nineteenth- and early-twentieth-century spatial models—will splinter into other Anthropocene sites. Clearly, they will not simply replicate past or present spaces. The polis, medieval walled cities, nineteenth-century industrial cities: each had their own historically unique spatial characteristics, subjectivities, modes of governance, and relations with nature. So too, Anthropocene spaces—managerial, predatory, and liberatory alike. Located as it is at the front lines of sea rise and branded as an experimental adaptation model for cities worldwide, perhaps Miami offers an image of the future: Miami, capital of the Anthropocene, as Los Angeles was for Edward Soja and Allen Scott a trendsetter marking out a post-Fordist trajectory subsequently followed by cities globally?[110] Or, given Miami's unique geology and the fragmenting, "pluriversal" rather than homogeneous, nature of life in the Anthropocene, perhaps Miami is more accurately auditioning one potential urbicidal pathway.[111] Instead of the image of Earth viewed from space, crisscrossed by interlinked commodity, industrial resource, or transportation networks—which so well illustrate contemporary global urbanization—here the paradigmatic visual would be those lit-up networks blinking out, as they are severed and parts of the world go island.

5

ANTHROPOCENE CRITICAL URBAN THEORY AND PRACTICE
RESEARCH AND DEVELOPMENT FOR A THAWING WORLD

Sometimes he wondered what zone of transit he himself was entering, sure that his own withdrawal was symptomatic not of a dormant schizophrenia, but of a careful preparation for a radically new environment, with its own internal landscape and logic, where old categories of thought would merely be an encumbrance.
—J. G. Ballard, *The Drowned World*

It is . . . vital to understand that what half-worked for the 1950s will not be adequate for the qualitatively different issues to be fought over the nature of civilization in the twenty-first century.
—David Harvey, "Cities or Urbanization?"

The City at the "End"

The end of the city is not new. For geographer Sarah Knuth, this trope is endemic to modern urban regimes, a fact that can be seen in the myriad targeted interventions into urban form and longevity.[1] Austerity, deindustrialization, and ghettoization, like gentrification or touristification, target and destroy urban geographies.[2] From creative destruction to "ruination-as-usual,"

Knuth describes capitalist urban regimes' propensity for devaluation and ruination in terms of Marx's classic: "all that is solid melts into air." "Eschatological evocation[s] of urban apocalypse," as Matthew Gandy puts it, are as old as the city itself, and feature prominently in Anthropocene imaginaries.[3] Megacities like Lagos, for example, are frequently characterized hyperbolically by military strategists as images of the crowded, chaotic century to come.[4] Apocalyptic imaginaries likewise have a history in Miami. In the late 1970s and early 1980s the city was labeled "Murder City, U.S.A., where crime had gone 'berserk'" and cocaine trafficking reigned.[5]

But the Anthropocene is also producing distinct visions of the urban's end, resilience and postresilience adaptation imaginaries that are novel both for Miami and cities more generally. In the past, as in resilient urban future visions today, even if technocrats bulldozed neighborhoods or narcotics-trade-driven violence prompted elite anxieties, the city itself—its basic structures and identity—persisted, despite being radically transformed or disrupted. As historian Gregory Bush puts it of what he calls Miami's "low point . . . after the McDuffie riots and the Mariel boatlift in 1980, when Time Magazine's cover article featured the area as 'Paradise Lost'":

> Had the promotional bubble burst with the exposure of racial brutality in the police department, broken promises to the African American problems that had been so long overlooked by the white power structure? Hardly. There was always another hastily devised "Action Plan" to thwart grievances and promote the area amid uncertain economic times.[6]

Carrying this pattern forward into a hoped-for eternity of expansive planetary urbanization, too often urban resilience approaches today lay bare what has always been the aim of urban governments and real estate interests: to recalibrate management techniques so as to preserve and conserve order, profit, and existing urban systems, even throughout crisis conditions, whether these are floods, hurricanes, or spectacular media imaginaries of urban climate apocalypse. After all, the same *Time* magazine that labeled Miami "Paradise Lost" in 1980, as discussed in chapter 1, declared the city the "capital of cool" in 2022.

But what we have seen is that, in Miami, a new, different "end of the city" is now being imagined. Rather than secure existing land formations and urban boundaries against sea rise, as resilience approaches seek to do, postresilience islandization visions discussed in the previous chapter in one way or another entail preemptive dismantling of an iconic-but-seen-as-doomed modern urban form and its infrastructural linkages. Here, despite the apparent affinity between urban development, resilience, and traditional forms of urbicide, we see what is nevertheless distinct about Anthropocenic end-of-the-city imaginaries, which envision not just planned undoing of this or that site, community, or neighborhood but of much of the city's entire built environment itself—as well as the infrastructural networks linking it to global commodity and production chains. Pushing already-existing islandization imaginaries to extreme, here Anthropocene island imaginaries also envision an attendant break from the globally networked energy and production infrastructures that constitute contemporary planetary urbanization. In these imagined islands, all that which is solid—not just buildings, but Miami's existing urban form itself and the infrastructural networks that subtend it—does not melt into air. Instead, the city's physical ground becomes the raw material for creating novel geographical forms. These imagined spatial territories are not intended to serve urban centers elsewhere but to sustain themselves.

In exploring the Anthropocene island imaginaries in chapter 4—which are but some of many variations on the theme discussed in the city—my aim has been not only to document a new, as-yet-unanalyzed urban imaginary emerging in Miami in response to the perceived hard limits of climate change, but also, in doing so, to reconsider taken-for-granted assumptions and imaginaries in urban scholarship and practice regarding the apparently endless resilience of existing cities and globalized urban systems in the Anthropocene. "Our long love affair with the city has reached new heights. It will only continue and intensify," writes Brendan Gleeson; "'planetary' really charts the final frontier, the telos of any earthly spatial fix," explains urban theorist Andy Merrifield, and regarding *alternatives* to contemporary formations of planetary urbanization, critical urban scholarship most often imagines only a new configuration or ways of managing the existing city— "the right to the city," for example, as many put it.[7] Certainly, there is nothing wrong with such imaginaries and efforts. However, my claim is that, though hypothetical—and by no means dominant—South Florida's end-of-the-city imaginaries raise important questions for critical urban thinkers and practitioners focused on planetary urbanization, urban resilience, and climate change, particularly those concerned with charting contemporary mutations in urbanized space and processes. Most obviously, they suggest the possibility that the "'lifetime' of today's urban fabrics and metropolitan forms," as Knuth puts it, may be more ephemeral than accounts of the Anthropocene and global urbanization as necessarily companion processes imagine. The twenty-first century's changing environments—and projections of them—and technopolitical adaptive responses may well lead to the end of seemingly unquestionable spatial forms like the urban or globally networked urbanization, and birth other, previously unimagined geographies.

As a possible trajectory alongside or in conflict with urban resilience, an urbicidal Anthropocene might encompass the brutal destructive agency of nonhuman forces. Already, record heat has buckled roads in the northwestern United States, and melting permafrost in northern Russian cities has crumbled buildings.[8] Very recently, the potential becoming-unviable of urban ground has painfully come to the fore in Miami, with media, residents, and engineers investigating the possible role of sea rise and saltwater corrosion (alongside shoddy construction, structural defects, and misleading government assurances) in the tragic Champlain Towers South condo collapse in June 2021.[9] A reproblematizing is already underway of the city's older buildings, with several now declared unsafe and evacuated, potentially paving the way for developer demolitions.[10] Or, Anthropocene urbicide might also issue from human agents—of all types and with manifold intentions—dismantling "already-obsolete" modern forms as fundamental as the urban itself, to use them as raw materials for new spatial typologies.[11] Such efforts may well be launched in *anticipation*—rather than in response to actual occurrence—of climate change impacts. And any such developments in urban environments would be multiple in form. But what we saw in chapter 4 was that at least one potential end of the urban might be tied to attempted delinking from infrastructural networks, rather than the eco-cybernetic interlinking that contemporary global urbanization and urban resilience both envision and promise.

Urban Theory/Practice for a Thawing World

The ends of the city explored in chapter 4 are, at present, just thought experiments. Nevertheless, allowing them to momentarily suspend the sense of inevitability that surrounds urban resilience in the "epoch of the human" opens thought to the possibility that both cities as we know them and current historical-political globally integrated networks may not be the only fate of human life in the Anthropocene. Likewise, the resilient subject, defined and shaped by vulnerability and survival and enrolled to inhabit this resilient city, need not be the end point of human subjectivity. In this opening, many questions become important for critical urban thought and practice. What extreme twenty-first-century spatial forms and processes may emerge, *within* urbanization, or to *disrupt, rework, unmake,* or *replace* urbanization? What new modes of government and forms of life might such geographies produce? Is it possible to imagine desirable modes of urban experimentation and dismantling beyond or against current urban structures, rather than leaving these to always emerge reactively as environmental adaptations or managerial interventions?

As I argued in the introduction, addressing these kinds of

questions requires a new research agenda: Anthropocene critical urban theory/practice. To begin, Anthropocene critical urban theory/practice can engage more closely with the spatial imaginaries, strategies, and forms of climate change adaptation emerging in specific places and contexts. Failure to do so risks reproducing the problem of relying on inherited spatial concepts—even newer ones like urban resilience, planetary urbanization, or responses to urban concepts "under stress"—and missing mutations that climate adaptations may produce, including nascent spaces and processes that push beyond currently dominant paradigms of urban resilience and globally interconnected urbanization.[12] Just as there is a need for conceptual, methodological, and practical experimentation to let go of what Neil Brenner calls "inherited cognitive maps of the urban condition," there is also a need to maintain an experimental openness to the possibility that urban resilience is not the only condition of life in the Anthropocene.[13] This is of course not a new problem. As geographer David Harvey, presaging Brenner's challenge, stated in the 1990s of the difficulty of developing adequate concepts to describe current urban contexts, "I worry that last year's conceptual tools and goals will be used to fight next year's issues in a dynamic situation that more and more requires proactive rather than remedial action."[14] But arriving at one's own content and imaginaries has today become more important than ever. To understand the Anthropocene's seismic mutations—both human- and nature-induced—deductive analyses based on conceptual frameworks and imaginaries from the twentieth and early twenty-first centuries will likely be inadequate on their own. Fresh ideas and imaginaries will be needed to grasp—and influence—novel spatial forms and extreme strategies produced at the intersections of urbanization, climate change, and adaptation, especially as these are also shaped by the existential and epistemological thawing that the Anthropocene also names.

Anthropocene critical urban theory/practice is therefore in part a matter of imagination. Imagination is not only a tool or matter of strategy, as governance experiments view and harness it. Rather, as Simin Davoudi reminds us, imagination is a matter of freedom.[15] This includes the freedom to abandon entrenched frameworks, ideologies, and imaginaries; to dream and actually make a future defined neither by floating co-work centers nor by preparedness hubs; and to take up profound imaginal and practical experimentation on one's own terms. It concerns equally the ability to envision a life beyond resilient survival in the Anthropocene, especially as the environmental, political, and social structures of industrial liberal society unravel. As we saw in chapters 2 and 3, to preserve urban systems amid climate-change-related shocks and stresses, governments, planners, and artists are forwarding a new vision of the city as an integrated social-ecological-technological system and, with it, enframing humans and nature alike as critical infrastructure. There is nothing inherently new or wrong with designing nature for human uses, or with the actual adaptive practices of human communities that are now called infrastructure. However, as we have seen, current definitions of nature and people as interlinked infrastructure are historically specific *governing* imaginaries, cybernetic management strategies oriented toward preservation of existing systems amid turbulence. If urban thinkers and practitioners tether their own imaginations and practice to this existing and limited governmental lens, their ability to freely imagine and strategize their own destinies beyond the limits of survivability will remain foreclosed. Stripped of the fundamentally human capacities for free imagination, they risk becoming subject to dominant imaginaries and thereby unable to stop the Anthropocene from progressing in its current, radically uneven shape. Not only that. With each reiteration of resilience logics and assumptions, they contribute *to* this imaginary, helping give it form and texture. Imaginaries are, after all, not just something done *to* individuals; as Davoudi puts it, "they are repertoires in which everyone plays their part."[16]

In light of present socio-ecological upheavals and governmental recalibrations launched in their name, it is important to subject the ideas, imaginaries, and real artifacts that Anthropocene critical thought and practice are producing to critical examination, and to analyze how these mirror or reinforce dominant resilience imaginaries. Here it is also worth considering what critical thought and practice are themselves at risk of becoming: in the same way that natures and human adaptive capacities are now rendered as eco-cybernetic feedback mechanisms to better optimize urban governance, it is also design, planning, science, and disciplines of all kinds that, in reproducing resilience's imaginary of human life on Earth, risk becoming cybernetic feedback, providing fuel for already-dominant imaginaries and systems of resilience. In the same way that ecosystems and human communities are enrolled as "living" critical infrastructure, human creative and critical activity is at risk of becoming *imaginal infrastructure,* part of a broader critical life-support system for the repetition of existing regimes. The effect of this is to limit the transformative potential of the present moment, to reign thought and action back within existing limits precisely at a moment when those limits need to be overcome.

With inherited ways of thinking and seeing wearing thin, how, as philosopher Chiara Bottici asks, are we to "operate a reactivation of the productive imagination, a reactivation of a space where drives and desires can just break free?"[17] For this, what is needed is not "more imagination" per se—such calls are ubiquitous already and overwhelmingly serve to reinforce dominant resilience imaginaries, with which the imagination of so many disciplines is already spellbound—but instead strategies through which the imagination can operate, beyond and without the footholds of this imaginary in all its iterations. Needed in all disciplines and fields of practice are ways to forge such a capacity able to think and research beyond dominant imaginaries, and to operate autonomously according to one's own styles and interpretations of the urban spaces and processes churned up by this epoch. Doing so requires taking risks, including the risk of breaking with the present order and its authorized imaginaries. Instead of unquestioningly repeating or celebrating urban resilience narratives that are ubiquitous with urban planning, design, and critical thought—allowing these to describe for us how we experience and live the Anthropocene—critical thinkers might ask whether such framings themselves pose a field of adversity of their own. Doing so may lead us into conflict with them, or simply to create other concepts and practices. While resilience embraces imagination, what it cannot imagine is a world that cannot be optimized or synthesized into its feedback systems. But this is ultimately both the problem and the solution: to borrow the poet William Blake's often cited words, "the Imagination is not a State: it is the Human Existence itself."[18]

In this book I have tried to think problematically about a series of climate adaptation projects and imaginaries as they are coming together in Miami's unique historical, political, economic, and environmental contexts. As political theorist Carol Bacchi puts it, using a problem-based approach of this kind—rather than assuming everything is the same everywhere, relying on transcendent or inherited guides—"produces a kind of freedom."[19] Thinking problematically about the Anthropocene-urban nexus frees us from a unitary interpretation of it as a static, inevitable "thing"—to use the term Ben Anderson has employed to describe what resilience is not—instead allowing us to analyze urban resilience as one situated, strategic response to a certain historical problematization of the present.[20] So too for islandization and urban ecological security zones, and even the urban condition itself as we know it today. More importantly, it also frees us to consider other possible trajectories, including our own understandings of the problems we confront. In the Anthropocene, old frameworks are coming undone, and re-creating modes of thinking and living is of paramount importance. Miami's effort to be an imaginal and infrastructural test lab is a reminder that powerful forces are already shaping the future of urban life on this terrain. Engaging this process as critical urban thinkers and practitioners can mean taking up critique in a traditional sense as a matter of unveiling power's hidden logics, as well as exploring the problematizations to which it is tied as political in their own right. Instead of accepting hegemonic visions of resilience and its moderate parameters as the only legitimate response to the Anthropocene, this could include exploring possibilities even within resilience's own frameworks.[21] But more broadly, it can also mean deciding for ourselves, in our own places and ways, what counts as a problem in the first place, how it is defined, what adversaries or limits we ourselves perceive, and how we choose to respond to them. After all, the point of critical urban theory/practice should not only be to understand government, but also to explore the wide range of ways in which humans problematize their lives and, as Foucault has put it, "give form to their behavior."[22]

The Human or the World?

One morning as I ordered coffee at Manolo, a popular Argentine bakery and restaurant in the working-class North Beach neighborhood of Miami Beach, a scene unfolded that captured something essential about Miami. An old but very large aquarium in the center of the restaurant came apart at the seams. The tank, holding perhaps as many as one hundred gallons of water, suddenly burst, sending water streaming to the floor. Panicked waiters and a fish deliveryman desperately held the aquarium back from complete collapse: one waiter used a belt to hold three sides of the aquarium together while another waiter and the fishmonger used their bare hands to clamp the edge. In spite of the failing aquarium flooding the floor, sunbaked customers were as unconcerned as the fish in the tank and went on eating their breakfast as if this were just another Miami oddity. In some ways they were right that the whole scene was Miami in miniature: chaotic, constantly breaking down, and often held together by jury-rigged solutions. But what this image depicts is also that the workers really cared. They sprinted for buckets from the fish purveyor down the street, and they used their collective strength to hold the sides of the tank together, desperately trying to prevent the full-on deluge of the restaurant. In contrast to some municipal authorities, for whom resilience is a necessary annoyance (or impediment) to their immediate grifts and gains—or simply a necessary antidote to climate crisis imaginaries—here one sees that ordinary working-class Miamians would do all they can to make a viable life in the city. However, in Miami as in all other cities, there is an extreme asymmetry between the capacity of the very rich to experiment with detachment from Anthropocene crises and systems and that of the ordinary urban dweller to do the same.

Chapter 2 discussed the sharp increase in popularity of efforts to build local infrastructures or learn survival skills. While often driven by a desire to retake control over one's means of existence, too often these scattered efforts become disaster preparedness for living in a biopolitical regime where the definition and horizon of life has already been reduced to survival. Such projects typically employ rudimentary resources and materials: recycled barrels for rainwater harvesting, handmade bow drills for firemaking, drinking water desalination using crumpled old Poland Springs bottles, and so forth. It is instructive to place these alongside the scale, vision, and material means of market-driven, top-down, or enclaved Anthropocenic islands described in chapter 4. Miami-based firm Arkup's "future proof" livable yacht, for example, costs $5.5 million and promises the means to "live in comfort and luxury in total autonomy, and enjoy life between the sea, the sky and the city" (Figure 16).[23]

Too often, critical thinkers dismiss such projects as the sole purview of elites or as inherently negative schemes of stubborn modern humans. After all, it is increasingly the shared view of many critical thinkers that human life in the Anthropocene should be humbled, that hubristic agency was the source of today's crises. In this vein, rather than providing alternatives to dominant imaginaries, too often critical academic taking up of the Anthropocene has generated a similar imaginary of human life to that of resilience. This "Anthropocenic thinking," to use David Farrier's term, with its emphasis on the power of the nonhuman world and rendering of humans as always already entangled in it, imports resilience's tonalities of precarity and permanent crisis while pushing further to reject even the

Figure 16. The Arkup 75 livable yacht, a floating luxury home designed to be "future proof": off-grid, solar-powered, rainwater-capturing, and able to field extreme weather events. Copyright: Arkup, Avant-Garde Life on Water, arkup.com.

hubristic agential will that underpins it.[24] Such an emphasis on chastening the modern agential subject and the hubris that undergirded diverse modern projects and instead imagining the ideal Anthropocene subject as one defined by humility in the face of the nonhuman world, precariously entangled in its flows and processes, is widespread among critical thinkers. Reversing the modern story of human freedom as a matter of rising above, separating from, or audaciously trying to transform the world, this line of Anthropocene thinking argues its inverse: that subjection to volatile Earth forces and entanglement amid ruins and crisis are the real nature of human existence. This entanglement in social-ecological-technological relations, once elided by modern thinking, is not to be escaped but to be embraced, argues philosopher Timothy Morton; such loops, he maintains, are humanity's "fate" and "destiny."[25] Seeing ourselves within complex systems to which humans are bound reveals that rather than makers of worlds, we are defined by precarity. Instead of holding on to hope or dreams of a "happy ending," humans must now learn to survive, to use anthropologist Anna Tsing's terminology, "in capitalist ruins."[26] This imaginary of life in the Anthropocene, political theorist David Chandler argues, is itself "increasingly affirmed as a positive and enabling opportunity" to overcome modernist frameworks and embrace the actual nature of life: precarious survival amid entangled crises and ruins.[27]

Such visions of human life in the Anthropocene resonate with and even radicalize those of life as survival for the poor and working classes who are often embedded in dominant resilient city imaginaries. Notably, Anthropocenic thinking's prescriptions also resonate with recent governmental shifts toward discourses and measures of green austerity—"the end of abundance, the end of recklessness," as French president Emmanuel Macron put it in the context of soaring energy costs, inflation, and cascading crises[28]—which is clearly a nascent climate dispositif in need of more extensive critical analysis. This emerging eco-austerity regime is not just about economic austerity but is an attempt to prepare and produce a particular *reality* for the twenty-first-century Anthropocene, and to delineate who is permitted to continue enjoying or practicing certain elements of human existence and Earth—and in what ways—and who is not.

Anthropocenic thinking recognizes something important about the present moment: we require new ways of thinking and living. But the task of thought should never be to instruct others in how to live, to provide universals to define life, or to flank hegemonic assertions that "nothing else is possible" other than the resilient continuation of existing structures while the world burns and floods. Taking seriously the claim that the old world is ending opens up much broader horizons. The end of the "one world world"—or what I have called the Anthropocene's "back loop"—is not the time for reasserting new universal definitions for what life should be, but for reaching out into the infinite range of what we and others might make it. Just as scholars such as Clive Barnett and Stephen Collier have argued against reductive or ontologizing critiques of government, so too is there a need for critical Anthropocenic thinking to resist this approach.[29]

Acknowledging that the world is not there "for us," that the Earth has its own vital forces and autonomy, need not require as its corollary the disavowal of human capacities. Surely there are other possibilities beyond this binary that some versions of Anthropocenic thinking tether human being to. As Chandler suggests, "perhaps it is a false and forced choice to choose between 'the human' and 'the world'? Perhaps rethinking modernity does not necessarily involve the refutation of any possibility of political alternatives other than those based on accepting our newfound fragility and vulnerability?"[30]

Questioning recently inherited Anthropocenic imaginaries and theoretical orientations can lead thinkers and practitioners to exciting new research perspectives, questions, and pathways. And, returning to the urban context, perhaps doing so might lead us to ask why urban design, planning, and life *need* be limited to the small, modest, and rudimentary, the unresourced and precarious? Does not Arkup's vision of "liv[ing] in comfort and luxury in total autonomy, and enjoy[ing] life between the sea, the sky and the city"[31] speak to the needs and dreams of so many urban dwellers—to *escape* constant manufactured precarity and crisis—expressed in the widely embraced turn to preparedness and survival skills discussed in chapter 2? Does Arkup's vision and the aesthetic and material means through which it is realized not capture some of the potentials and horizons of what human imaginative and adaptive capacity could become in the twenty-first century—and not just that of a few? This vision of serenity and peace—not crisis and survival—speaks to a type of security increasingly imagined to be impossible for many in the Anthropocene, but one worth exploring now in ever-new ways, conceptually and materially, and at scales both small and large, via technologies both high and low.[32]

Cosmotechnics at/for the "End"

The Anthropocene calls for the reappropriation of imagination in a double sense: freeing it from the parameters of resilience, but also freeing urban scholars, designers, planners, governments, engineers, artists, scientists, and residents alike to explore its use as a tool for researching and developing the material structures and processes needed for extreme transformation—at all scales. Given the high stakes of the Anthropocene's political, social, and environmental challenges, it will not only be new imaginaries that are crucial but equally the exploration of material strategies to make other ways of living possible. And so, not limited to tracing elite experimentation or salvage-oriented epistemologies and practices, Anthropocene critical urban theory/practice can also apply a free, upended mindset to considerations of how urban thinkers and practitioners intervene in, resist, or reappropriate Anthropocenic urban mutations rather than being subject to them. Here again, freedom and autonomy to question and, when needed, jettison even recently inherited ways of thinking are essential.

Resilience, adaptation, and survival are often conflated. But as the Alliance of the Southern Triangle emphasize in their "Protocols for the Phase Transition," discussed in the previous chapter, design for adaptation to transforming environments is not necessarily just about survival and resilience. Nor is it exclusively a matter of humans abdicating their agency to a vastly more powerful nonhuman world of forces and flows, accommodating themselves to its agency and demands alone. Rather, they write, "it is within our synthetic capacities that our greatest potential emerges."[33] Here the accent is actually on the human, and in this vein we can draw inspiration from geographer Nigel Clark's work.[34] While readers tend to approach his writings with the volume up on the nonhuman, Clark's work is remarkable for the endless stories of the awesome—creative, destructive, evasive—things *humans* do on and with a planet of powerful forces and matter. Clark gives us account after account of human-created flame and fire, shelter and warmth, defense and offense.[35] Drawing on a favorite form of matter—fire—he describes how humans across place and time have deployed, captured, and intensified flame for manifold uses and ends both instrumental and irrational: to illuminate dark places; blaze trails through tangled landscapes; create hunting grounds, ward off predators, or drive prey; bake figurines and breads; burn fossil fuels, power factories; create warmth and comfort; choreograph elaborate fireworks displays; produce weapons, armor, and money; or torch structures, for example during war or uprisings.

Fire is of course but one of the many strategies through which very different humans, across place and time, in political upheaval and everyday life, take hold of and "tap the power and potentiality of our planet . . . [the] elemental ingredients of our inherited worlds . . . the material-energetic expressiveness of our planet itself"[36] and put them to different ends. Some of these are survival strategies, others a means to dominate other people or lands. But human existence past and present is also replete with creative, destructive, and evasive design strategies through which humans take hold of and shape, intensify, or destroy environments and matter as a means of life—or liberation from tyranny. These strategies work on and through a variety of elements: earth, fire, water, swamps . . .[37]

In any of these, instead of a force to which humans must only accommodate themselves or submit, the power of elemental and Earth processes can be understood as an incitement, a provocation, a gift, or, just as equally, a challenge or threat. Engaging in such designerly activity, humans make aspects of the world quite knowable: how to start a fire, how to contain and carry it, and so forth. But such strategies just as equally are waged within contexts of real, often existential uncertainty. Fire also moves of its own accord; an accidental spark blowing on the wind can set whole forests and towns ablaze. But far from eliminating possibilities of human agency or political action, this uncertainty is the dynamic context within which human action is waged. From this vantage point, that many of Earth's forces and elements are unknown or uncontrollable constitutes not grounds for self-hatred but part of the beauty and tragedy of life, a context within and against which deeply varied, not to mention hubristic, human strategies are forged.

This image of human life in but also able to separate itself from and act on the world diverges radically from both resilience and Anthropocenic thinking approaches to the twenty-first century. In contrast to homogeneous visions of an endlessly controlled and governed human and nonhuman world—a total order whose completion would be achieved in the form of eco-cybernetic planetary urbanization, perfected—this is a view of a wild and burning planet populated by wild and burning, but also creative and strategic, people as well, who, rather than being subject to the world, shape and push back against it in various ways.[38] Exploring the potentialities of these and so many other points of "innervating contact with the elements and powers of the earth"[39] illustrates how irreducible to dominant governing imaginaries human life on Earth is. Such strategies are not just objects of remembrance—missed opportunities or replicas behind museum glass—but human forms that can be studied and elaborated in ever-new ways.

Anthropocene critical urban theory/practice can take this complex, profoundly beautiful understanding of human-environment adaptive design in hand and use it as a lens through which to research the vast range of synthetic strategies through which diverse actors are transforming, and can transform, specific Anthropocene territories and socio-ecological contexts. Through this exploration, new conceptual frameworks for understanding lifeways, imaginaries, and human-environment relationships will emerge. After all, again following philosopher Yuk Hui, such activities as experimentation with infrastructure for water, food, housing, energy, and beyond are never just technical solutions, but are a *cosmotechnical* matter of worldmaking, joining of earth and sky.[40] In Hui's thought, cosmotechnics

> is the unification of the cosmos and the moral through technical activities, whether craft-making or art-making. There hasn't been one or two technics, but many cosmotechnics. What kind of morality, which and whose cosmos, and how to unite them vary from one culture to another according to different dynamics. I am convinced that in order to confront the crisis that is before us—namely, the Anthropocene, or the intrusion of Gaia (Latour and Stengers), or the "entropocene"

(Stiegler), all presented as the inevitable future of humanity—it is necessary to reopen the question of technology, in order to envisage the *bifurcation* of technological futures by conceiving different cosmotechnics.[41]

What Hui suggests, and what Anthropocene critical urban theory/practice can embrace, is not only envisioning possibilities for life that outstrip dominant resilience imaginaries, but also a widespread, interdisciplinary exploration of the cosmotechnics of food, shelter, communication, transportation, and energy needed to break away from current locked-in trajectories. This would therefore be an experimental research-and-development agenda—resonant with that envisioned by the island imaginaries discussed in chapter 4—at once technical and imaginal, conceptual and material, of the "technodiversity" seen as needed to reinvent and revitalize human life on Earth. Hui describes technodiversity as "a variety of ways of understanding and constructing technology in different cultures, emerging from the locality and always in exchange with other localities," and I agree with his statement that "the quest and promotion of technodiversity should be one of the core tasks for us today."[42] That the exploring of the material means of existence has become so widespread and that so many ordinary urban dwellers are seeking material answers suggests that, along with shrinking biodiversity and stormier weather, the Anthropocene may also be marked by a powerful potential to rediscover and reclaim the human capacity to dwell on Earth, via cosmotechnical means.

Living with Water, Reconsidered

South Florida has a rich history of human adaptations to what others deem unlivable environments, a great number of which are essential not just to survival but what practitioners consider cultural and political autonomy (and, more broadly, the good life). For example, the Miccosukee's combining of techniques for inhabiting the Everglades and resisting assimilation underscore this human ecological connection between design for living with aquatic environments and autonomy. The Miccosukee were originally part of the Hitchiti-speaking Lower Creek Nation in Alabama and Georgia. In the 1800s, with the Indian Wars, many Miccosukee were relocated west or killed. However, many others refused surrender and instead escaped to the Everglades. The river of grass became a refuge for them, allowing survivors to resist assimilation and relocation. To preserve their spatial and cultural autonomy, they developed practices suited to the Everglades environment, including living in small camps atop hammocks, which are small, elevated habitats dispersed across the ecosystem. On these they constructed chickees, which are open-air elevated homes made of cypress poles and sabal palm frond or palmetto thatched roof. Traveling through the watery environments via hollowed-out cypress trunk logs, propelled by pole or sail, they fished largemouth bass, gar, and turtles. With their once staple crop, corn, almost impossible to grow, they developed techniques of harvesting cabbage palm that grew native in the raised hammocks.[43] The Miccosukee prioritization of autonomy through human-environment relations persists to the present day. Their biannual weeklong water study is a community-wide Everglades monitoring event that brings Miccosukee students, tribal elders, and families out on large caravans of swamp buggies and ATVs to test the water's dissolved oxygen levels, turbidity, and salinity and discuss Miccosukee culture. The sampling event is just as much a way to maintain independent control over water quality, fight back against polluting runoff from commercial agriculture in the region, and preserve the health of the ecosystem. It is also but one example of how the Miccosukee use science and technology alongside traditional knowledge to fight to maintain their autonomy against state jurisdiction, cultural assimilation, and industrial pollution, in and with the environment. As Miccosukee artist, poet, and environmentalist Houston Cypress frequently emphasizes, the need to respect Indigenous sovereignty is key for the tribe, and they insist on the freedom to handle their affairs and culture by and for themselves.[44] Notably, toward this end, the Miccosukee are conducting experimental efforts to construct their own elevated tree islands in the Everglades to prepare for flooding—whether from sea-level rise or that produced through restoration-induced rehydration.[45]

In a different register, Miami is home to unique lifeways and styles of vernacular architecture particular to its hot, wet environment and that are also inextricably tied to sovereignty. For example, the city has long been home to a culture of houseboats, stilted and floating homes, with some residents even putting prefab homes or shipping containers on barges or Styrofoam and concrete bases to float independently in the area.[46] The city once had a large houseboat community, and for many, part of the attraction to the floating lifestyle was the freedom it offered and the comradery generated among houseboaters. In a different vein, we also might think of the underground communications networks, food provisions, and maritime infrastructure and know-how that made up the "saltwater railroad," through which slaves in Florida and the American South made nautical escape by boats to the nearby Bahama Islands.[47] Or, as architect Germane Barnes highlights, Miami is home to a unique, environmentally specific style of historic Black architecture found in the early houses built in neighborhoods such as Coconut Grove. This unique "littoral urbanism," he writes, "found in the Deep South . . . requires a significant commitment to the sustainability of local ecologies. . . . Raised floors and porches, central corridors, and passive cooling are all components of historic black Miami architecture that contemporary Miami has rebranded as climate resiliency and environmentally-friendly design. Unfortunately, today many of the historic architecture remnants are no longer present, and what is left is government-sanctioned housing developments."[48] Or, finally, we might think of the work of anthropologist Laura Ogden, whose monumental books *Swamp Life: People, Gators, and Mangroves Entangled in the Everglades* and *Gladesmen: Gator Hunters, Moonshiners, and Skiffers* collect Gladesmen diaries and lifeways for living with humor on their own terms in the hot and buggy Everglades. Central to Ogden's account is the story of the Bill Ashley gang, a 1920s band of outlaws whose close knowledge of the Everglades landscape enabled them to evade capture by hiding in the swamps for years.[49]

Today, South Florida's unique and diverse water-based cultural and architectural character is, much as Barnes laments, largely being lost as legal restrictions clamp down on floating cultures, suburban and luxury development replaces vernacular architectures and lifeways, and even the diverse histories of living with water in the region are being renarrated as the impetus for urban resilience strategies. Nevertheless, many of these cultures persist. And, assessing Miami's present and future prospects, nothing seems more called for than an interdisciplinary investigation into such techniques and cultures past, present, and for the future. Not merely for building resilience or surviving promised crisis conditions, but equally for rewilding the unique character and texture of the region, regenerating the human and environmental landscape that defines it: a renaissance of extreme, Anthropocenic vernacular architectures, skills, infrastructures, and cultures.

Like Miami, all cities rest on vast reservoirs of freedom-oriented adaptations and techniques for handling and manipulating changing environments. But the research-and-development agenda I am proposing need not focus merely on reproducing models from the past. Often, both Anthropocene theorists and survival skills instructors employ discourses of returning to traditional or "lost" skills that assume an unchanging environment. But skills, strategies, and cosmotechnics themselves are already being, we saw in chapter 2, reinvented, in and for twenty-first-century urbanized and extreme environments of all kinds. Along these lines, chapter 3 discussed the threat sea-level rise is seen to pose to Miami's fresh drinking water supply, a potential existential crisis that has led some commentators to predict the city's demise due to saltwater infiltration into aquifers and water shortages. However, as hydrogeologist Virginia Walsh points out, "there will always be drinking water here. It's just a question of how much you want to pay for it."[50] If ordinary residents of Miami or other water-threatened cities such as Atlanta intend to avoid the fate imagined in novelist Paolo Bacigalupi's *The Water Knife*—a near-future vision of Phoenix "killed" by climate-change-induced drinking water shortages, where corporations and governments control the Colorado River's little remaining water, the very rich live in lush, closed-loop arcologies with flush toilets and running water, and the common people crowd together under the scorching heat awaiting Red Cross water rations, recycling urine to quench their thirst—then it makes sense to explore other, autonomous infrastructural options. As in the Caribbean Islands, there are traditions and techniques of rainwater harvesting in the Florida Keys that beg to be explored anew and updated. Ultimately, whether invasive python hunting or large-scale desalination plants, in South Florida as in all other regions, the question becomes, what skills, what tools, what strategies for inhabiting Anthropocene environments on one's own terms—and how to break away from currently locked-in trajectories?

These local considerations raise a much broader and essential point. In chapter 2 we saw how urban resilience discourses and practices often reserve the bird's-eye view and massive, transformative planning for government while relegating the ordinary urban dweller to surviving within disastrous conditions. Everything about this partitioning, as well as the critical perspectives on the bird's-eye view itself, needs to be questioned. Future scenario visioning workshops such as those discussed in chapter 2 are valuable not only because they open imagination but also because they call us to think strategically about the future. A great deal becomes possible when people do this together, beginning from their own realities (rather than the abstract, fictitious "we" that governs each and all). But, from there, it is not necessary to limit first-person sight to building individual reactive adaptive capacity, or to abdicate the bird's-eye planner's view to government. Scholars have long critiqued the mastery that inheres in the bird's-eye view, a distance that Gillian Rose, for example, sees as requiring a desire to control and dominate.[51] Within Anthropocene critical thinking, hubristic planning is similarly condemned as an outdated relic of modern humanist agency, with theorists such as Tsing arguing that the Anthropocene requires relinquishing intentionality.[52] But in the Anthropocene it is more necessary than ever to continuously question recently received truisms just as much as spatial concepts and imaginaries, particularly when they stand in the way of transformation. After all, there is nothing inherently wrong with a bird's-eye view or with planning. Both are simply how humans step back and look at their worlds in order to shape them. In this vein, Jasper Bernes suggests that we distinguish between central planning and decentral planning. In the former, he reminds us, systems of surveillance and control will always be required to enforce compliance of the subjects on which it is imposed. However, he continues,

> Central planning does not exhaust the meanings of planning as such. Planting a field is planning, as is, looking a bit further out, the rotation of crops. Scattering seeds in an area to which one's tribe will return in six months is planning. The forests that European explorers encountered in North America had, in fact, been made into big-game hunting parks by the controlled, planned use of fire to clear undergrowth. Irrigation canals and viaducts, the Great Wall and Machu Picchu: all planning.[53]

Reappropriating resilience's own tools—such as futures scenario visioning or even "social infrastructure"—may offer some means to get out of the structured precarity that defines life for ever-greater numbers of urban thinkers and practitioners. The futures that stand to emerge from such a reappropriation on the ground in Miami or any other city could point beyond the resilience of currently dominant socioeconomic urban systems—those that produce structural "insecurity by design"[54]—and toward other very different pathways. After all, in Clark's redefinition, much like adaptive strategies discussed above, envisioning future scenarios is not just a means of maintaining the status quo but is fundamentally "an exploration of potentialities that inhere in the Earth and cosmos." As Clark emphasizes, "even under conditions of intense environmental pressure, there is always an excess of possibility nestling in the strata and flows of the Earth."[55]

With this in mind, the contemporary reclamation of imaginative and cosmotechnical powers could point toward the gestation of new spaces and lifeways, rather than acceptance of those that now exist as the only ones possible. Rather than a future defined by floating co-work centers or one of unending disaster survival, harnessing strategic future-oriented thinking—and the extreme infrastructures and skills it often brings into view as necessary—has the potential to open the way for actualizing the transformed, utopian futures imagined in the best of scenarios. What new urban spaces, processes, and designs could this redirection produce? What rich territories of culture and anthropotechnical invention, sociality and human-environment interaction, could be woven?

Urban Depresencing

Beginning and end are often impossible to disentwine. While human agency in the Anthropocene is nearly always theorized as a matter of continuation and survival—design for the resilience of urban spaces; recalibration of new technomodernist systems to allow systems to continue; or arts of collaborative survival amid neoliberalism's ruinous infrastructures and environment—chapter 4 noted that designers are increasingly considering other, more extreme possibilities for human intervention, specifically those oriented toward dismantling. Many critical scholars might view the urbicidal imaginaries explored there as destined only to produce what Mike Davis and Daniel Bertrand Monk call neoliberal "evil paradises."[56] But, given urban resilience's often-conservative, don't-rock-the-boat-too-much nature, geared toward maintaining existing socioeconomic structures, perhaps there is something compelling in the willingness to envision scrapping one of the country's most economically polarized cities and building something different. After all, this is precisely the transformative potential in the otherwise negative Anthropocene: the possibility of undoing existing systems, stopping the annihilation of human and nonhuman life-forms, and experimenting with other ways of living. Not because disaster is coming, but because it is already here, not least in the form of the already-degraded and homogenized built and natural environments that surround and shape human life today.

This raises complex questions answerable only by the people affected by them. But together with urban experimentation and creation of new ways of living, could there be positive, democratic versions of Anthropocene urban disassembly, waged not by technocrats or developers upon helpless populations but by designers, planners, and populations against imposed urban regimes or projects or rotting, "zombie" cities that have themselves become hostile to human life?[57] Would this even be desirable? Such questions have been raised by philosopher Alexandre Monnin, who argues that today we inherit a "negative commons" of toxic soils, power plants, abandoned oil wells, and perpetually leaking nuclear disaster sites like Fukushima.[58] These already-obsolete processes and infrastructures, Monnin argues, are driving the Anthropocene and environmental devastation, and therefore cannot just be abandoned, made "use" of, or lived with. This is especially so in cities and urbanized environments.

The point is not that making new use of such systems is impossible. Think for example of the people of Bougainville's repurposing of the ruins of a Rio Tinto Zinc corporation mine and local island ecologies to make weapons, instruments, electrical grids, medicine, and forms of spirituality to sustain themselves and successfully fight off the Papua New Guinea army and the RTZ corporation's plundering of the island's resources. This example suggests powerful scales and scopes of repurposing far beyond the small life-support systems and lifeways often associated with Anthropocene culture and art. Still, Monnin argues, it is difficult to imagine a power plant or abandoned oil well rendered "inoperative" "by liberating the potentials that have remained inactive in it in order to allow a different use of them"—as Giorgio Agamben's thinking around "use" puts it.[59]

Recalling Brooks + Scarpa's Unplanning Miami proposal discussed in chapter 4, Monnin and his colleagues Emmanuel Bonnet and Diego Landivar's consideration of the Anthropocene leads to the positing of a conceptual and pragmatic challenge for designers, engineers, and communities: closing already-obsolete processes and unusable, environmentally destructive systems. Design, they argue, has historically been a matter of bringing to presence, that is, "a practice of creation oriented towards the future and underpinned by an 'ameliorative' intention."[60] However, such a task—often framed today in terms of maintenance, sustainability, or resilience—fails to recognize the extent to which design's foundations have been tested by the Anthropocene. Given the "negative commons" we inherit, the task of design, they argue, can no longer be the "recovery of the world," but instead resides in the careful, considered dismantling of the systems that compose it.[61] This perspective departs from notions of "the commons" as that which is beneficial to a community and in need of proper management or reappropriation, such as those associated with Elinor Ostrom or Antonio Negri.[62] It also departs from calls for "dwelling in the ruins" or "staying with the trouble," such as those often associated with Anthropocenic thinking.[63] Instead, the commons envisioned here is an inherited hostile environment and wasteland that must be depresenced in order for life to thrive. Rather than dwelling in ruins, for Monnin, "the good life from now on is the one that will take charge of these realities and work to dismantle them."[64]

If Earth is to be made habitable—not just survivable—then instead of merely maintaining what is, Monnin argues for a material approach focused on "closing worlds."[65] In contrast to oft-repeated calls to build new worlds in the Anthropocene, Monnin and colleagues foresee "the dawn of a world to be 'deprojected.'"[66] To describe this activity they propose the concept of "destauration"—the reverse of philosopher Étienne Souriau's "instauration"—which they define as "the action of diminishing the existence of a being, deintensifying it."[67] In marked contrast to much thinking around the notion of "degrowth," Monnin emphasizes the nonmoralizing, cosmotechnical, and poetic nature of destauration. These depresencing strategies and arts, he and colleagues argue, require "learn[ing] to dis-innovate, to decommission and dismantle infrastructures incompatible with the ecological situation."[68] Of course, they emphasize, there are complicated technical and ethical questions associated with doing so, which are answerable only by those materially engaged with them. In their view, arts of depresencing must be explored by those on the ground concerned with and affected by the infrastructure or system in question. What does this mean, how does it work, who decides, and what are the cosmotechnical tools involved? Nevertheless, in the urban context, such questions pose important challenges to dominant modes of thought regarding resilience and represent important research pathways for Anthropocene critical theory/practice to explore.

"A Conscious Experimentation in Urban Form"

The questions raised in this book are but a few of the matters that rethinking the resilient urban Anthropocene's inevitable dominance opens up. Exploring them will create important new pathways of engagement for critical urban thinkers and practitioners and require new conceptual and practical tools appropriate to a thawing world. Ultimately, critical urban thought and practice for the Anthropocene need not limit itself to a choice between intellectual or practical, the imaginative or the technical, but can be part of a democratic—diffuse, in situ—experimental research agenda able to question and shed already-obsolete ideas and forms including, if need be, those as seemingly inevitable as the urban itself.

Whether oriented toward preservation, bringing to presence, or depresencing, urban imaginaries and strategies for the Anthropocene will come from the places and people they concern. A key point that living in Miami in the Anthropocene drives home is that attempts to produce art, cajole political action, or even just describe urban life today that attempt to apply rote frameworks to complex realities will inevitably fail. Similarly, there is no pure space from which tools or material strategies will be invented. As Miami curator and writer Gean Moreno emphasizes of life in the city, subjectivity there, now, is not a coherent or bounded thing. Understanding the city is not a matter of choosing one pure narrative reality over another—whether that of journalist and media narratives of impending climate apocalypse and technocratic governmental responses; tech happy hours and crypto-trading delirium; nor this or that culture or traditional knowledge. Instead, it requires thinking the collision of each of these and other often-contradictory realities, and understanding that there is not *one* city but many, a multivalent Anthropocene city that folds in complex local socioecological legacies and contemporary realities of race and class, infrastructure and imagination, and points toward multiple, potentially contradictory trajectories still to be determined. Rather than from preordained practices or ideologies, it is from *within* this fraught "glitch" space, as Moreno calls it, that the designs to make any other future will have to be dreamed and developed.[69]

This point is relevant to Anthropocene critical urban theory/practice more broadly. Glitch subjectivity is a particularly apt description of living at Miami's feverish intersection of submerged future projections and "capital of cool" status. But, as the epistemological and political codes that grounded modern Western societies break down, glitch subjectivity also becomes a far more widespread condition. This is not necessarily something to lament or celebrate, but a complex, thawing terrain to inhabit. As resilience ecologists emphasize, when socio-ecological systems break down, freedom increases and predictability declines. What forms of urban thought and imagination, design and strategy can come from inhabiting *this* place—one of contradictions and emergent formations, which *is* the Anthropocene?

Anthropocene critical urban theory/practice is not intended for a specific, existing political ideology or a movement based on a specific political organization or identity. Just as I have proposed unmooring from inherited imaginaries and strategies, given the environmental and epistemological upheavals of our time perhaps it is equally important to free urban thought and practice to operate without the banisters of existing political ideologies. The Anthropocene critical urban theory/practice proposal is for a mode of thought and action that does not interpret and act in the world based on inherited (or assumed external or transcendent) ideologies, but instead begins from problems and questions unfolding there where it is enacted. The intention of phrases like "freedom-oriented adaptations" is to indicate this—not to signal a specific libertarian politics. Anthropocene critical urban theory/practice is a research-and-development agenda to be taken up by anyone interested in its questions, provocations, and approach: urban scholars, theorists, designers, planners, engineers, scientists, artists, policymakers, water managers, and residents alike, anyone concerned with thinking the present and reopening the future.

According to David Graeber and David Wengrow, recent archaeological evidence suggests that early cities were not, as in some popular portrayals, always hierarchical forms of human habitation defined by the presence of a ruling elite, bureaucrats, central planners, and police.[70] Contrary to popular mythology, they argue, large agglomerations of people living together at scale does not inherently lead to bureaucracy and class exploitation. Instead, sifting through recent archaeological studies of early cities, including Teotihuacan and recently discovered Taljanky in Ukraine, estimated to be as old as 4100 B.C.E., shows the presence of manifold varieties of urban agglomerations with no evidence of palaces or monuments to elites, but instead variegated forms of more egalitarian cohabitation and autonomous organization of urban life. Emphasizing the diversity of innovation found across so many early cities, they conclude: "It's not just that some early cities lack class divisions, wealth monopolies, or hierarchies of administration. They exhibit such extreme variability as to imply, from the very beginning, a *conscious experimentation in urban form.*"[71] Abandoning unquestioning allegiance to the imperative of resilience as the only future for human life in the Anthropocene frees urban thought and practice to again take up this age-old human adventure.

CODA
ZONE OF TRANSIT

Domingo Castillo's 2017 film *Tropical Malaise: Prologue* premiered as a 7,000-square-foot projection in Miami Beach, on landscape architecture firm West 8's Soundscape Park, next to the Frank Gehry–designed New World Center building.[1] The film depicts Miami Beach 500 million years in the future. No humans, no buildings, no roads, no traffic. Just crashing waves and wind-blown trees. Set to a hammering electronic soundtrack, the film takes its viewer VR-style across an undulating aquatic planet. At first glance, the film offers an image of the now-standard vision of Miami's endgame: a future without us. But Castillo—who is perhaps Anthropocene Miami's most vital artist—is showing us something more: the *production* of such imaginaries of the city's end. The short film opens with a designer-side view of tutorial videos for Maxon Cinema 4D, a digital rendering software used to produce real estate plans, AR/MR/VR, or movies like *Avengers: Endgame* and *Pacific Rim*. A designer's cursor manipulates the program interface, dragging and pasting, adjusting object properties—ocean resolution, ocean size, wind speed, choppiness, ocean depth, foam threshold—to create images of a churning, peopleless, aqua-blue ocean (Figure 17).

Instead of another cinematic, pedagogical forecast of Miami's impending doom, *Tropical Malaise* is a portrait of how Anthropocenic imaginaries, including those submerged future visions that

| 199

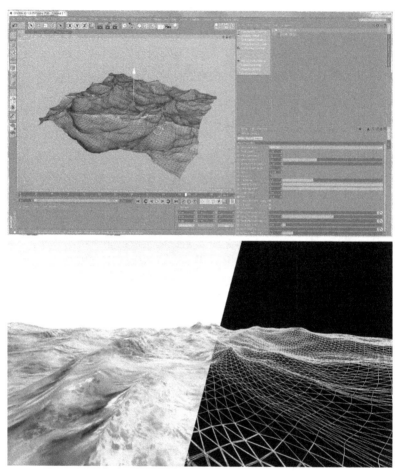

Figure 17. Stills from Domingo Castillo's *Tropical Malaise: Prologue*, which features a backend view of sea-level-rise inundation generated through digital rendering software. Courtesy of Domingo Castillo (see endnote 1 for details about the film).

are so ubiquitous and powerful in the city, are being produced here and now. On one hand, their imagined future is constructed and locked in materially by Miami's business-as-usual development. As writer and curator Natalia Zuluaga explains, the urban planning, architecture, and real estate renderings generated by the software that codes the waves in Castillo's film "have helped propel the phenomenon of the global city, and the *Disneyfication* of place, both of which allow the idea of the city—and all of the capital infrastructure that supports it—to be prototyped and pasted on real terrain in real time."[2] Structures like New World Center are real-world manifestations of this, but such materialities aren't even necessary to generate real effects in Miami. Indeed, argues writer and curator Gean Moreno,

> The only consequential visual production generated in and about Miami at the moment takes the form of promotional architectural animations. Even if produced for investors and potential clients, these artifacts endure long lives on different media platforms and, in this way, structure an image of the city that, while independent of some realities on the ground, is in no way unreal. The money these objects move, the desires that they awaken, and the effects of their existence on urban reality (to say nothing of our subjectivities) are quite concrete.[3]

One could say much the same of the visual and discursive problematization of Miami as climate change ground zero and a doomed future Atlantis. Just as the waves of *Tropical Malaise* are neither natural nor real but rather the effects of code, neither are the submerged Miami of interactive digital sea-rise maps nor the resilient Miami of municipal planning documents. Rather, through their discourses, narratives, and visualizations, these and other imaginaries sketch, code, and write certain visions of Miami's fate. As Zuluaga writes of architectural designs, "they render reality by projecting from the future and bearing down on the present in very real ways. These renderings represent blackbox processes wherein desire, power, and speculation coagulate and become visible. They are moving images that motivate power and capital to catalyze into reality."[4] The same is true for visions of Miami's resilience or its end, which are produced as images and narratives repeat over and over until a certain urban future telos is cemented, internalized, and accepted as true.

In reality, urban futures are still less predictable than this. At present the dominant response is that of maintaining existing systems. However, many other possibilities exist for urban life in the Anthropocene, limited only by the range of diverse responses that will be launched to its social, environmental, and political-economic dislocations. What may emerge is not one city but many ways of living in the city—or far beyond it—some of which may coexist while others may lead to conflict over the future and nature of urbanized life.

Still, one thing is critical to keep in sight. What Jefferson Cowie and Joseph Heathcott wrote of deindustrialization's revelation of the "ephemeral quality of the world we take for granted" also applies to Miami or any other coastal city: because capital is currently fixed in high-rise condos and mansions, and because asphalt and complex water management systems order the parameters of life in the city, the particular cultural-technological urban configuration we call Miami exudes "an aura of permanence, durability, and heritage."[5] But the starchitect-designed high-rises, gated private art collections, ghettoized neighborhoods, and even the iconic imported palm trees are all, in fact, but "artifacts" of a particular historical, spatial, economic conjuncture in which capital touched down here, in this place at this time. And the feeling of eternity this conjuncture radiates "mask[s] a fundamental impermanence; it obscure[s] the forces that both created this world through investment" and can "[break] it apart by withdrawing investment."[6] In the same way, the "end" of Miami as we know it may not be from water or human adaptive intervention but rather from a real estate crash or withdrawal of investment and insurance. Once again, however, possibilities don't actually end there. At the southernmost tip of the United States, standing at the ocean's edge, feet lapped by salt water and skin kissed by endless sun, one must admit that the city is alive. The City of Miami Beach spends millions of dollars on maintenance to keep the beautiful, pristine order of the city's palms, sending trucks out every day to pick up stray fallen fronds. But nature continues to grow uncontrollably of its own accord. Iguanas sunbathe alongside roads and on tabletops. The sky is constantly shifting, morphing with each passing stormfront. At the end of the day no one knows what is going to happen to Miami.

Rather than accept clichéd stories as the only future possible, a critical urban theoretical/practical approach adequate to the Anthropocene requires that we continually ask, What lies beyond the limits of dominant imaginaries? What else is possible in the intertidal? Will the technological breakthroughs gestured to by politicians come to be, allowing Miami's continued existence for another hundred years? Or, will the housing bubble burst, leading businesses to divest in favor of other cities, and ordinary residents to create other futures in their absence? What extreme architectures and territories might such post-resilience experimentation produce?

Ultimately, considering such questions requires setting aside inherited imaginaries and exploring the possibilities beckoning now, even if they are just seething below the pristine manicured surface, even if they disturb comfortable assumptions. What starts to appear possible—conceptually and materially—when one inhabits the strange "zone of transit" into which the Anthropocene propels us, rather than tethering oneself to Holocenic refuges of thought and action?

ACKNOWLEDGMENTS

Research for and writing of this book were generously supported by an Urban Studies Foundation Postdoctoral Research Fellowship. I am deeply grateful to the USF, and especially want to thank Donald McNeill, Desiree Fields, Joe Shaw, and all the USF Trustees. At Florida International University I want to express my gratitude to Kevin Grove for his extensive and continual support, and also want to thank Melissa Bernardo, Jody Grove, Guillermo Grenier, Ulrich Oslender, Gail Hollander, and John Stuart. Research for this book was also supported by the Florida Coastal Everglades Long-Term Ecological Research Program. Much of my understanding of Everglades history and hydrology comes from my colleagues in the FCE network, and I am grateful to Mahadev Bhat, Jessica Cattelino, Evelyn Gaiser, John Kominoski, Jayantha Obeysekera, Laura Ogden, Randall Parkinson, Rene Price, Mike Ross, and Tiffany Troxler for sharing their knowledge with me. Special thanks to my colleagues in the Department of Urban and Regional Planning at Florida Atlantic University: Eric Dumbaugh, Peter Henn, Serena Hoermann, Yanmei Li, Melina Matos, Louis Merlin, Diana Mitsova, John Renne, and Craig Snyder. Thank you also to Jeff Huber in FAU's School of Architecture, and to everyone at Life University. Appreciation always to my friends and collaborators in Miami and South Florida: Domingo Castillo, Caroline Castro, Houston

Cypress, Rodney Devera, Felice Grodin, Elite Kedan, Gean Moreno, Lena Moreno, Anastasia Samoylova, George Sánchez-Calderón, Jack Shealy, Barron Sherer, Misael Soto, and Natalia Zuluaga, as well as to El Espacio 23 for hosting me. I want to express my sincere gratitude to all those who took time to share their knowledge and perspectives on South Florida futures with me, especially Tom Gustafson, Fane Lozman, and Nicolas Derouin. Thank you also to Jeremy Salloum at HistoryMiami Museum for assistance locating images.

The ideas in this book have been shaped by conversations with and feedback generously provided by friends and interlocuters including but not limited to Kieran Aarons, Ross Exo Adams, Nikhil Anand, Alexander Arroyo, Niall Atkinson, Nick Axel, Daniel Barber, Grga Bašić, Bjørn Enge Bertelsen, Bruce Braun, Neil Brenner, Caroline Busta, Evan Carver, David Chandler, Jason Cons, Randi Elin Gressgård, Roxana Fabius, Arne Harms, Alysha Helmrich, Patricia Margarita Hernández, Lil Internet, David Iwaniec, Elizabeth Johnson, Daniel Keller, Ed Keller, Katrin Klingan, Lukas Ley, Suhail Malik, Lelani Mannetti, Alexandre Monnin, Katherine Nelson, Lavínia Pereira, Jonathan Pugh, Julian Reid, Idris Robinson, Rory Rowan, Martin Savransky, Judith Schelehe, Sabina Shaikh, Taylor Shelton, and Simone Tulumello, as well as the participants at the talks, conferences, and workshops where over the past several years I shared early versions of material in this book. Thank you as well to all my students, to Pieter Martin at University of Minnesota Press, and, above all else, to my family.

NOTES

Introduction

1. Will Steffen, Paul J. Crutzen, and John R. McNeill, "The Anthropocene: Are Humans Now Overwhelming the Great Forces of Nature?," *Ambio* 36, no. 8 (2007): 614.

2. Giorgio Agamben, *The Time That Remains: A Commentary on the Letter to the Romans* (Stanford: Stanford University Press, 2005), 23.

3. Paul Crutzen, "Geology of Mankind," *Nature* 415 (2002): 23; Jan Zalasiewicz, Colin N. Waters, Erle C. Ellis, Martin J. Head, Davor Vidas, Will Steffen, Julia Adeney Thomas, Eva Horn, Colin P. Summerhayes, Reinhold Leinfelder, J. R. McNeill, Agnieszka Gałuszka, Mark Williams, Anthony D. Barnosky, Daniel de B. Richter, Philip L. Gibbard, Jaia Syvitski, Catherine Jeandel, Alejandro Cearreta, Andrew B. Cundy, Ian J. Fairchild, Neil L. Rose, Juliana A. Ivar do Sul, William Shotyk, Simon Turner, Michael Wagreich, and Jens Zinke, "The Anthropocene: Comparing Its Meaning in Geology (Chronostratigraphy) with Conceptual Approaches Arising in Other Disciplines," *Earth's Future* 9, no. 3 (2021): e2020EF001896; Stephanie Wakefield, *Anthropocene Back Loop: Experimentation in Unsafe Operating Space* (London: Open Humanities Press, 2020).

4. Bernard Stiegler, *The Neganthropocene* (London: Open Humanities Press, 2018), 39.

5. Owen Gaffney and Will Steffen, "The Anthropocene Equation," *Anthropocene Review*, 4, no. 1 (2017): 53–61; Anna Tsing, *The Mushroom at the End of the World* (Princeton: Princeton University Press, 2016); Simon Dalby, *Anthropocene Geopolitics: Globalization, Security, Sustainability* (Ottawa: University of Ottawa Press, 2020); Wakefield,

Anthropocene Back Loop; Déborah Danowski and Eduardo Viveiros de Castro, *The Ends of the World* (Cambridge, UK: Polity Press, 2017).

6. Dalby, *Anthropocene Geopolitics;* David Chandler, *Ontopolitics in the Anthropocene: An Introduction to Mapping, Sensing, and Hacking* (London: Routledge, 2018).

7. Naomi Klein, *This Changes Everything: Capitalism vs. the Climate* (New York: Simon & Schuster, 2014); Tsing, *Mushroom;* Wakefield, *Anthropocene Back Loop;* Scott Hamilton, "Securing Ourselves from Ourselves? The Paradox of 'Entanglement' in the Anthropocene," *Crime, Law, and Social Change* 68 (2017): 579–95; Roy Scranton, *Learning to Die in the Anthropocene: Reflections on the End of a Civilization* (San Francisco: City Lights Publishers, 2015).

8. On the Anthropocene's disruption of modern ontologies and human/nature binaries, see, e.g., Tom Cohen, Claire Colebrook, and J. Hillis Miller, *Twilight of the Anthropocene Idols* (London: Open Humanities Press, 2016); Bruno Latour, *Facing Gaia: Eight Lectures on the New Climatic Regime* (Cambridge, UK: Polity Press, 2017); David Farrier, *Anthropocene Poetics* (Minneapolis: University of Minnesota Press, 2019). For critical work on the Anthropocene and its relation to historically specific forms of capitalist accumulation and control, see, e.g., Andreas Malm and Alf Hornborg, "The Geology of Mankind? A Critique of the Anthropocene Narrative," *Anthropocene Review* 1, no. 1 (2014): 62–69; Henrik Ernstson and Erik Swyngedouw, eds., *Urban Political Ecology in the Anthropo-Obscene: Interruptions and Possibilities* (Abingdon: Routledge, 2019); Christophe Bonneuil and Jean-Baptiste Fressoz, *The Shock of the Anthropocene: The Earth, History, and Us* (London: Verso, 2016).

9. Jason Moore, *Anthropocene or Capitalocene? Nature, History, and the Crisis of Capitalism* (Oakland, Calif.: PM Press, 2016); on Industrial Revolution–based periodization of the Anthropocene, see, e.g., Steffen, Crutzen, and McNeill, "The Anthropocene." On urban spaces as both Anthropocene driver and artifact, see, e.g., Hillary Angelo and David Wachsmuth, "Why Does Everyone Think Cities Can Save the Planet?," *Urban Studies* 57, no. 11 (2020): 2201–21; Roger Keil, "Paved Paradise: The Suburb as Chief Artefact of the Anthropocene and Terrain of New Political Performativities," in Ernstson and Swyngedouw, *Urban Political Ecology*, 165–83; Thomas Elmqvist, Erik Andersson, Timon McPhearson, Xuemei Bai, Luís Bettencourt, Eduardo Brondízio, Johan Colding, Gretchen Daily, Carl Folke, Nancy Grimm, Dagmar Haase, Daniel Ospina, Susan Parnell, Steve Polasky, Karen Seto, and Sander

Van der Leeuw, "Urbanization in and for the Anthropocene," *Urban Sustainability* 1, no. 1 (2021): article 6.

10. On the urban age as a demographic matter, see *United Nations Department of Economic and Social Affairs, Population Division, World Urbanization Prospects: The 2018 Revision* (ST/ESA/SER.A/420) (New York: United Nations, 2019); Ricky Burdett and Deyan Sudjic, eds. *The Endless City: The Urban Age Project by the London School of Economics and Deutsche Bank's Alfred Herrhausen Society* (London: Phaidon, 2007). On planetary urbanization, see Neil Brenner and Christian Schmid, "Towards a New Epistemology of the Urban?," *City* 19, nos. 2–3 (2015): 151–82; Andy Merrifield, *The Politics of the Encounter: Urban Theory and Protest under Planetary Urbanization* (Athens: University of Georgia Press, 2013); Brendan Gleeson, "Critical Commentary: The Urban Age," *Urban Studies* 49, no. 5 (2012): 931–43.

11. Jonathan Silver, "Suffocating Cities: Climate Change as Social-Ecological Violence," in Ernstson and Swyngedouw, *Urban Political Ecology*, 129–47; Elmqvist et al., "Urbanization in and for the Anthropocene."

12. Angelo and Wachsmuth, "Why Does Everyone?"; Sue Ruddick, "Situating the Anthropocene: Planetary Urbanization and the Anthropological Machine," *Urban Geography* 36, no. 8 (2015): 1113–30; Ihnji Jon, *Cities in the Anthropocene: New Ecology and Urban Politics* (London: Pluto Press, 2021).

13. Stephanie Wakefield and Bruce Braun, "Governing the Resilient City," *Environment and Planning D: Society and Space* 32, no. 1 (2014): 4–11; Bruce Braun, "A New Urban Dispositif? Governing Life in an Age of Climate Change," *Environment and Planning D: Society and Space* 32, no. 1 (2014): 49–64; Steward T. A. Pickett, Brian McGrath, M. L. Cadenasso, and Alexander J. Felson, "Ecological Resilience and Resilient Cities," *Building Research & Information* 42, no. 2 (2014): 143–57,

14. Erik Swyngedouw, "More-than-Human Constellations as Immuno-Biopolitical Fantasy in the Urbicene," *The New Geography* 1 (2017): 20–27. Franciszek Chwałczyk has similarly suggested the term "Urbanocene." Franciszek Chwałczyk, "Around the Anthropocene in Eighty Names—Considering the Urbanocene Proposition," *Sustainability* 12, no. 11 (2020): 4458–4533.

15. Jan Zalasiewicz, Mark Williams, Colin Waters, Anthony Barnosky, and Peter Haff, "The Technofossil Record of Humans," *Anthropocene Review* 1 (2014): 34–43.

16. Neil Brenner and Christian Schmid, "The 'Urban Age' in Question," *International Journal of Urban and Regional Research* 38, no. 3 (2014): 750.

17. Neil Brenner, "Theses on Urbanization," *Public Culture* 25, no. 1 (2013): 94; Neil Brenner, *Critique of Urbanization: Selected Essays* (Basel: Birkhäuser Verlag, 2016), 22.

18. Neil Brenner, *New Urban Spaces: Urban Theory and the Scale Question* (Oxford: Oxford University Press, 2019); Christian Schmid, Ozan Karaman, Naomi C Hanakata, Pascal Kallenberger, Anne Kockelkorn, Lindsay Sawyer, Monika Streule, and Kit Ping Wong, "Towards a New Vocabulary of Urbanisation Processes: A Comparative Approach," *Urban Studies* 55, no. 1 (2018): 19–52; Lauren Rickards, Brendan Gleeson, Mark Boyle, and Cian O'Callaghan, "Urban Studies after the Age of the City," *Urban Studies* 53, no. 8 (2016): 1523–41; Brenner and Schmid, "Towards a New Epistemology"; Neil Brenner, ed., *Implosions/Explosions: Towards a Study of Planetary Urbanization* (Berlin: Jovis, 2014); Teresa P. R. Caldeira, "Peripheral Urbanization: Autoconstruction, Transversal Logics, and Politics in Cities of the Global South," *Environment and Planning D: Society and Space* 35, no. 1 (2017): 3–20.

19. Swarnabh Ghosh and Ayan Meer, "Extended Urbanization and the Agrarian Question: Convergences, Divergences and Openings," *Urban Studies* 58, no. 6 (2021): 1097–1119; Martín Arboleda, *Planetary Mine: Territories of Extraction under Late Capitalism* (London: Verso, 2020); Brenner, "Theses on Urbanization"; Brenner and Schmid, "The 'Urban Age'"; Merrifield, *The Politics of the Encounter;* Rickards et al., "Urban Studies"; Neil Brenner and Nikos Katsikis, "Operational Landscapes: Hinterlands of the Anthropocene," *AD/Architectural Design* 90, no. 3 (2020): 22–31. See also work by geographers and urban political ecologists which critically analyzes urban-nature metabolisms and relational, more-than-human urban networks and systems. For an overview, see, e.g., Bruce Braun, "Environmental Issues: Writing a More-than-Human Urban Geography," *Progress in Human Geography* 29, no. 5 (2005): 635–50.

20. Wakefield, *Anthropocene Back Loop.*

21. Neil Brenner and Swarnabh Ghosh, "Between the Colossal and the Catastrophic: Planetary Urbanization and the Political Ecologies of Emergent Infectious Disease," *Environment and Planning A: Economy and Space* 54, no. 5 (2022): 867–910; Roger Keil and S. Harris Ali, "The Urban Political Pathology of Emerging Infectious Disease in the

Age of the Global City," in *Mobile Urbanism: Cities and Policymaking in the Global Age,* ed. Eugene McCann and Kevin Ward (Minneapolis: University of Minnesota Press, 2011), 123–46.

22. Arboleda, *Planetary Mine.*

23. Kian Goh, "Urbanising Climate Justice: Constructing Scales and Politicising Difference," *Cambridge Journal of Regions, Economy and Society* 13, no. 3 (2020): 561.

24. On urban resilience and economic inequality, see, e.g., Neil W. Adger, Ricardo Safra de Campos, Tasneem Siddiqui, and Lucy Szaboova, "Commentary: Inequality, Precarity, and Sustainable Ecosystems as Elements of Urban Resilience," *Urban Studies* 57, no. 7 (2020): 1588–95; Sara Meerow and Joshua P. Newell, "Urban Resilience for Whom, What, When, Where, and Why?," *Urban Geography* 40, no. 3 (2019): 309–29; Joshua Long and Jennifer L. Rice, "From Sustainable Urbanism to Climate Urbanism," *Urban Studies* 56, no. 5 (2019): 992–1008. On urban resilience and racialized exclusion, especially in Greater Miami, see, e.g., Kevin Grove, Allain Barnett, and Savannah Cox, "Designing Justice? Race and the Limits of Recognition in Greater Miami Resilience Planning," *Geoforum* 117 (2020): 134–43; Kevin Grove, Savannah Cox, and Allain Barnett, "Racializing Resilience: Assemblage, Critique, and Contested Futures in Greater Miami Resilience Planning," *Annals of the American Association of Geographers* 110, no. 5 (2020): 1613–30; Anne Bonds, "Refusing Resilience: The Racialization of Risk and Resilience," *Urban Geography* 39, no. 8 (2018): 1285–91. On urban resilience and neoliberalism, see, e.g., Tom Slater, *Shaking Up the City: Ignorance, Inequality, and the Urban Question* (Oakland: University of California Press, 2021); Aiden While and Mark Whitehead, "Cities, Urbanisation, and Climate Change," *Urban Studies* 50, no. 7 (2013): 1325–31. On urban resilience and governmentality, see, e.g., Braun, "A New Urban Dispositif?"; Wakefield and Braun, "Governing the Resilient City"; Stephanie Wakefield and Bruce Braun, "Oystertecture: Infrastructure, Profanation, and the Sacred Figure of the Human," in *Infrastructure, Environment, and Life in the Anthropocene,* ed. Kregg Hetherington (Durham: Duke University Press, 2019), 193–215; Stephanie Wakefield, "Urban Resilience as Critique: Problematizing Infrastructure in Post-Sandy New York City," *Political Geography* 79 (2020): 102148; Kate Derickson, "Urban Geography III: Anthropocene Urbanism," *Progress in Human Geography* 42, no. 3 (2018): 425–35; Maria Kaika, "'Don't Call Me Resilient Again!' The New Urban Agenda as Immunology . . . Or . . . What Happens When

Communities Refuse to be Vaccinated with 'Smart Cities' and Indicators," *Environment and Urbanization* 29, no. 1 (2017): 89–102.

25. Grove, Barnett, and Cox, "Designing Justice?"; Chiara Camponeschi, "Beyond Bounce Back: A Healing Justice and Trauma-Informed Approach to Urban Climate Resilience" (PhD thesis, University of Guelph, Canada, 2020).

26. On resilience and neoliberalism, see, e.g., David Chandler, *Resilience: The Governance of Complexity* (Abingdon: Routledge, 2014); Ashley Dawson, *Extreme Cities: The Perils and Promise of Urban Life in the Age of Climate Change* (New York: Verso, 2017); Jonathan Joseph, "Resilience as Embedded Neoliberalism: A Governmentality Approach," *Resilience: International Policies, Practices and Discourses* 1 (2013): 38–52. On resilience and government, see Brad Evans and Julian Reid, *Resilient Life: The Art of Living Dangerously* (London: Polity, 2014); David Chandler, Kevin Grove, and Stephanie Wakefield, eds., *Resilience in the Anthropocene: Governance and Politics at the End of the World* (Abingdon: Routledge, 2020); Kevin Grove, *Resilience* (Abingdon: Routledge, 2018). On biopolitics, see Christopher Zebrowski, "Governing the Network Society: A Biopolitical Critique of Resilience," *Political Perspectives* 3, no. 1 (2008): 1–41; Jeremy Walker and Melinda Cooper, "Genealogies of Resilience from Systems Ecology to the Political Economy of Crisis Adaptation," *Security Dialogue* 14, no. 2 (2011): 143–60; Frederic Neyrat, *The Unconstructable Earth: An Ecology of Separation* (New York: Fordham University Press, 2019); Mark Vardy and Mick Smith, "Resilience," *Environmental Humanities* 9, no. 1 (2017): 175–79; Mark Neocleous, "Resisting Resilience," *Radical Philosophy*, 178 (March/April 2013): 2–7. On multiplicity and contingency, see James Brassett and Nick Vaughan-Williams, "Security and the Performative Politics of Resilience: Critical Infrastructure Protection and Humanitarian Emergency Preparedness," *Security Dialogue* 46, no. 1 (2015): 32–50.

27. On urbanization as a process, see, e.g., David Harvey, "Cities or Urbanization?," *City* 1, nos. 1–2 (1996): 38–61.

28. Stephanie Wakefield, "Making Nature into Infrastructure: The Construction of Oysters as a Risk Management Solution in New York City," *Environment and Planning E: Nature and Space* 3, no. 3 (2020): 761–85; Kian Goh, "Flows in Formation: The Global–Urban Networks of Climate Change Adaptation," *Urban Studies* 57, no. 11 (2020): 2222–40.

29. Kasia Paprocki, "All That Is Solid Melts into the Bay: Antici-

patory Ruination and Climate Change Adaptation," *Antipode* 51, no. 1 (2019): 295–315; Kasia Paprocki, *Threatening Dystopias: The Global Politics of Climate Change Adaptation in Bangladesh* (Ithaca: Cornell University Press, 2021).

30. Kellie Zhao, "The East Side Coastal Resiliency Project Breaks Ground, but Opponents Aren't Backing Down," *The Architect's Paper*, May 28, 2021, https://www.archpaper.com/2021/05/east-side-coastal-resiliency-project-breaks-ground-opponents-arent-backing-down/.

31. Benjamin Bratton, *The Stack: On Software and Sovereignty* (Cambridge: MIT Press, 2016).

32. Brenner, "Theses on Urbanization," 95.

33. "Miami metropolitan area" and "Greater Miami" are the most commonly used terms to refer to the regional three-county urban agglomeration that includes Miami-Dade County, Broward County, and Palm Beach County. Within this, Miami-Dade County contains thirty-four incorporated municipalities, including those for which "Miami" is best known: the City of Miami Beach and the City of Miami. Throughout the book, unless otherwise specified, for readability's sake I will use the term "city" or "Miami" when speaking of Greater Miami and the metropolitan area. I will specify times when I am focused more narrowly on experimental projects being carried out in specific municipalities, as in chapter 2's discussion of City of Miami Beach resilience projects.

34. U.S. Census Bureau, "County Population Totals and Components of Change: 2020–2022," 2023, https://www.census.gov/data/tables/time-series/demo/popest/2020s-counties-total.html; Jan Nijman, "Rethinking Suburbia: A Case Study of Metropolitan Miami," *Environment and Planning A* 47 (2015): 71–72.

35. N. D. B. Connolly, *A World More Concrete: Real Estate and the Remaking of Jim Crow South Florida* (Chicago: University of Chicago Press, 2014).

36. Alejandro Portes and Ariel C. Armony, *The Global Edge: Miami in the 21st Century* (Oakland: University of California Press, 2018), 16.

37. Erika Spanger-Siegfried, Melanie Fitzpatrick, and Kristina Dahl, *Encroaching Tides: How Sea-Level Rise and Tidal Flooding Threaten US East and Gulf Coast Communities over the Next 30 Years* (Cambridge, Mass.: Union of Concerned Scientists, 2014); Shimon Wdowinski, Ronald Bray, Ben P. Kirtman, and Zhaohua Wu, "Increasing Flooding Hazard in Coastal Communities Due to Rising Sea Level: Case Study of Miami Beach, Florida," *Ocean & Coastal Man-*

agement 126 (2016): 1–8; Southeast Florida Regional Climate Change Compact Sea Level Rise Ad Hoc Work Group, *Unified Sea Level Rise Projection for Southeast Florida* (2019 update), https://southeastflorida climatecompact.org/wp-content/uploads/2020/04/Sea-Level-Rise-Proj ection-Guidance-Report_FINAL_02212020.pdf [hereafter cited as Compact, *Unified Sea Level Rise*].

38. Miami-Dade County, *A Risk-Based Approach to Septic Systems Vulnerable to Sea Level Rise* (Miami: Miami-Dade County, 2020).

39. Arnoldo Valle-Levinson, Andrea Dutton, and Jon Martin, "Spatial and Temporal Variability of Sea Level Rise Hot Spots over the Eastern United States," *Geophysical Research Letters* 44, no. 15 (2017): 7876–82.

40. Urban Land Institute, *Stormwater Management and Climate Adaptation Review: Urban Land Institute Advisory Services Panel Report* (Washington, D.C.: Urban Land Institute, 2018).

41. Intergovernmental Panel on Climate Change, *Climate Change 2022: Impacts, Adaptation and Vulnerability, Contribution of Working Group II to the Sixth Assessment Report of the Intergovernmental Panel on Climate Change*, ed. Hans-Otto Pörtner, Debra C. Roberts, Melinda M.B. Tignor, Elvira Poloczanska, Katja Mintenbeck, Andrés Alegría, Marlies Craig, Stefanie Langsdorf, Sina Löschke, Vincent Möller, Andrew Okem, and Bardhyl Rama (Cambridge, UK: Cambridge University Press, 2022).

42. Compact, *Unified Sea Level Rise*, 9. These numbers are projections for SLR above the mean sea level recorded in 2000 in Key West, Florida.

43. Figure of 380 high-tide flood events per year from Union of Concerned Scientists, *Encroaching Tides in Miami-Dade County, Florida. Fact Sheet: Sea Level Rise and Tidal Flooding along the Atlantic Coast* (Cambridge, Mass.: Union of Concerned Scientists, 2016), 1; "permanent inundation" from Compact, *Unified Sea Level Rise*, 31.

44. Susan Hanson, Robert Nicholls, N. Ranger, Stéphane Hallegatte, Jan Corfee-Morlot, C. Herweijer, and Jean Chateau, "A Global Ranking of Port Cities with High Exposure to Climate Extremes," *Climatic Change* 104, no. 1 (2011): 89–111; Daniel Raimi, Amelia Keyes, and Cora Kingdom, "Florida Climate Outlook: Assessing Physical and Economic Impacts through 2040," Report 20–01 (Washington, D.C.: Resources for the Future, 2020).

45. Raimi, Keyes, and Kingdom, "Florida Climate Outlook."

46. Urban Land Institute, *The Business Case for Resilience in*

Southeast Florida: Regional Economic Benefits of Climate Adaptation (Washington, D.C.: Urban Land Institute, 2020), 6.

47. Frederick Bloetscher, Barry Heimlich, and Daniel E. Meeroff, "Development of an Adaptation Toolbox to Protect Southeast Florida Water Supplies from Climate Change," *Environmental Reviews* 19 (December 2011): 397–417.

48. Kate Stein, "'We're a Living Laboratory': Miami Beach Works on Resiliency as Businesses Face Flooding," *WLRN*, October 17, 2016, https://www.wlrn.org/news/2016-10-17/were-a-living-laboratory-miami-beach-works-on-resiliency-as-businesses-face-flooding.

49. Tristram Korten, "In Florida, Officials Ban Term 'Climate Change,'" *Miami Herald*, March 8, 2015, https://www.miamiherald.com/news/state/florida/article12983720.html.

50. Greater Miami and the Beaches, *Resilient 305*, May 30, 2019, https://www.mbrisingabove.com/wp-content/uploads/Resilient305_final.pdf.

51. U.S. Economic Development Association, "Biden-Harris Administration Designates Tech Hub in Miami, Florida to Advance Technological Innovation and Address the Global Climate Crisis," press release, October 23, 2023, https://www.eda.gov/news/press-release/2023/10/23/South-Florida-Climate-Resilience-Tech-Hub.

52. Travis Dagenais, "GSD Introduces the Future of the American City with Support from Miami's Knight Foundation," Harvard Graduate School of Design, March 20, 2018, https://www.gsd.harvard.edu/2018/03/gsd-introduces-the-future-of-the-american-city-with-support-from-miamis-knight-foundation/

53. C. S. Holling, "Resilience and Stability of Ecological Systems," *Annual Review of Ecology and Systematics* 4 (1973): 1.

54. Grove, *Resilience*; J. Walker and Cooper, "Genealogies"; Carl Folke, "Resilience: The Emergence of a Perspective for Social–Ecological Systems Analyses," *Global Environmental Change* 16, no. 3 (2006): 256; Lance Gunderson and C. S. Holling, eds., *Panarchy: Understanding Transformations in Systems of Humans and Nature* (Washington, D.C.: Island Press, 2002); Brian Walker, C. S. Holling, Stephen Carpenter, and Ann Kinzig, "Adaptability and Transformability in Social-Ecological Systems," *Ecology and Society* 9, no. 2 (2004): 5.

55. B. Walker et al., "Adaptability and Transformability," 4; see also Gunderson and Holling, *Panarchy*; Steward T. A. Pickett and J. M. Grove, "Urban Ecosystems: What Would Tansley Do?," *Urban Ecosystems* 12 (2009): 1–8; Fikret Berkes and Carl Folke, eds., *Link-*

ing *Social and Ecological Systems: Management Practices and Social Mechanisms for Building Resilience* (Cambridge, UK: Cambridge University Press, 1998).

56. Carl Folke, "Resilience (Republished)," *Ecology and Society* 21, no. 4 (2016): 44; Lance H. Gunderson, Craig R. Allen, and C. S. Holling, eds., *Foundations of Ecological Resilience* (Washington, D.C.: Island Press, 2009); Berkes and Folke, *Linking Social and Ecological Systems*; Folke, "Resilience: The Emergence"; Brian Walker, Lance Gunderson, Ann Kinzig, Carl Folke, Steve Carpenter, and Lisen Schultz, "A Handful of Heuristics and Some Propositions for Understanding Resilience in Social-Ecological Systems," *Ecology and Society* 11, no. 1 (2006): 13; Jianguo Liu, Thomas Dietz, Stephen R. Carpenter, Marina Alberti, Carl Folke, Emilio Moran, Alice N. Pell, Peter Deadman, Timothy Kratz, Jane Lubchenco, Elinor Ostrom, Zhiyun Ouyang, William Provencher, Charles L. Redman, Stephen H. Schneider, and William W. Taylor, "Complexity of Coupled Human and Natural Systems," *Science* 317 (2007): 1513–16.

57. Folke, "Resilience: The Emergence," 261; Gunderson and Holling, *Panarchy*.

58. Brenner, *Implosions/Explosions*, 21.

59. See, e.g., Gunderson and Holling, *Panarchy*.

60. Orit Halpern, Jesse LeCavalier, Nerea Calvillo, and Wolfgang Pietsch, "Test-Bed Urbanism," *Public Culture* 25, no. 2 (2013): 272–306.

61. As Ben Anderson argues, resilience is not a homogeneous "thing," but instead assembles heterogeneous tactics and place-specific designs. Anderson, "What Kind of Thing Is Resilience?," *Politics* 35, no. 1 (2015): 60–66; see also Arianna Tozzi, "An Approach to Pluralizing Socio-natural Resilience through Assemblages," *Progress in Human Geography* 45, no. 5 (2021): 1083–1104.

62. Grove, *Resilience*; Braun, "A New Urban Dispositif?"

63. Michel Foucault, *Power/Knowledge: Selected Interviews and Other Writings, 1972–1977*, ed. Colin Gordon (New York: Pantheon Books, 1980), 194.

64. Stephen Collier, "Topologies of Power: Foucault's Analysis of Political Government beyond Governmentality," *Theory, Culture & Society* 26, no. 6 (2009): 78–108; Stephen Collier, "Neoliberalism and Rule by Experts," in *Assembling Neoliberalism: Expertise, Practices, Subjects*, ed. Vaughan Higgins and Wendy Larner (New York: Palgrave Macmillan, 2017), 23–44. For a problem-based approach to resilience, see Philip Boyle, "'Building a Safe and Resilient Canada': Resilience

and the Mechanopolitics of Critical Infrastructure," *Resilience: International Policies, Practices, and Discourses* 7, no. 1 (2018): 59–82.

65. Clive Barnett and Gary Bridge, "The Situations of Urban Inquiry: Thinking Problematically about the City," *International Journal of Urban and Regional Research* 40, no. 6 (2016): 1186–1204; Michel Foucault, "The Confession of the Flesh," in *Power/Knowledge*, 194–228; Michel Foucault, "Polemics, Politics, Problematizations: An Interview with Michel Foucault," in *Essential Works of Foucault (Vol. 1)*, ed. Paul Rabinow (New York: The New Press, 1998), 111–19.

66. Clive Barnett, "On Problematization: Elaborations on a Theme in Late Foucault," *Nonsite* 16 (Summer 2015), https://nonsite.org/article/on-problematization; Colin Koopman, *Genealogy as Critique: Foucault and the Problems of Modernity* (Bloomington: Indiana University Press, 2013); Thomas Osborne, "What Is a Problem?," *History of the Human Sciences* 16, no. 4 (2003): 1–17; Paul Rabinow, "Dewey and Foucault: What's the Problem?," *Foucault Studies* 11 (2011): 11–19.

67. Carol Bacchi, "Why Study Problematizations? Making Politics Visible," *Open Journal of Political Science* 2, no. 1 (2012): 1–8; Roger Deacon, "Theory as Practice: Foucault's Concept of Problematization," *Telos* 118 (2000): 127–42.

68. Foucault, "Polemics, Politics, Problematizations."

69. Barnett, "On Problematization"; Clive Barnett, *The Priority of Injustice: Locating Democracy in Critical Theory* (Athens: University of Georgia Press, 2017); Nikolas Rose and Peter Miller, "Political Power beyond the State: Problematics of Government," *British Journal of Sociology* 43, no. 2 (1992): 173–205.

70. Foucault, "Confession of the Flesh," 195.

71. Koopman, *Genealogy as Critique*.

72. Michel Foucault, "What Is Enlightenment?," in *The Foucault Reader*, ed. Paul Rabinow (New York: Pantheon Books, 1984), 50.

73. Foucault, "Polemics, Politics, Problematizations," 118; Michel Foucault, *Fearless Speech* (New York: Semiotext(e), 2001), 171.

74. Foucault, *Fearless Speech*.

75. Paul Rabinow, *Anthropos Today: Reflections on Modern Equipment* (Princeton: Princeton University Press, 2003).

76. Foucault, *Fearless Speech*.

77. Bruce Braun, "Taking Earth Forces Seriously," in *Viscosity—Mobilizing Materialities*, ed. Karen Lutsky, Ozayr Saloojee, and Emily Eliza Scott (University of Minnesota Department of Architecture, 2019), 47–61. On the spatiotemporal approach to theorizing cities, see

also Eric Sheppard, Vinay Gidwani, Michael Goldman, Helga Leitner, Ananya Roy, and Anant Maringanti, "Introduction: Urban Revolutions in the Age of Global Urbanism," *Urban Studies* 52, no. 11 (2015): 1947–61.

78. Grove, Cox, and Barnett, "Racializing Resilience."

79. David Harvey, *The Urbanization of Capital: Studies in the History and Theory of Capitalist Urbanization* (Baltimore: Johns Hopkins University Press, 1985).

80. Barney Warf, "Splintering Urbanism: Networked Infrastructures, Technological Mobilities, and the Urban Condition," *Annals of the Association of American Geographers* 93, no. 1 (2003): 246–47; Ashley Carse, "Keyword: Infrastructure: How a Humble French Engineering Term Shaped the Modern World," in *Infrastructures and Social Complexity: A Companion*, ed. Penny Harvey, Casper Bruun Jensen, and Atsuro Mori (Abingdon: Routledge, 2017), 27–39; Nikhil Anand, *Hydraulic City: Water and the Infrastructures of Citizenship in Mumbai* (Durham: Duke University Press, 2017); Keller Easterling, *Extrastatecraft: The Power of Infrastructure Space* (New York: Verso, 2014); Colin McFarlane and Jonathan Rutherford, "Political Infrastructures: Governing and Experiencing the Fabric of the City," *International Journal of Urban and Regional Research* 32, no. 2 (2008): 363–74; Stephanie Wakefield, "Infrastructures of Liberal Life: From Modernity and Progress to Resilience and Ruins," *Geography Compass* 12, no. 7 (2018): e12377; Bruce Braun and Sarah Whatmore, *Political Matter: Technoscience, Democracy and Public Life* (Minneapolis: University of Minnesota Press, 2010); Stephen Graham and Simon Marvin, *Splintering Urbanism: Networked Infrastructures, Technological Mobilities and the Urban Condition* (London: Routledge, 2001); Erik Swyngedouw, *Social Power and the Urbanization of Water: Flows of Power* (Oxford: Oxford University Press, 2004).

81. Stephen Graham, *Disrupted Cities: When Infrastructure Fails* (London: Routledge, 2009); Braun, "Environmental Issues"; Ian Cook, "Follow the Thing: Papaya," *Antipode* 36 (2004): 624–64; Alex Hughes and Suzanne Reimer, eds., *Geographies of Commodity Chains* (London: Routledge, 2004); Deborah Cowen, *The Deadly Life of Logistics: Mapping Violence in Global Trade* (Minneapolis: University of Minnesota Press, 2014), among others.

82. Susan Star, "The Ethnography of Infrastructure," *American Behavioral Scientist* 43 (1999): 377–91; Brian Larkin, "The Politics and Poetics of Infrastructure," *Annual Review of Anthropology* 42 (2013): 327–43; Brian Larkin, *Signal and Noise: Media, Infrastructure, and*

Urban Culture in Nigeria (Durham: Duke University Press, 2008); Kregg Hetherington, "Surveying the Future Perfect: Anthropology, Development and the Promise of Infrastructure," in *Infrastructures and Social Complexity: A Companion*, ed. Penny Harvey, Casper Bruun Jensen, and Atsuro Mori (Abingdon: Routledge, 2017), 40–50; Kregg Hetherington, ed., *Infrastructure, Environment, and Life in the Anthropocene* (Durham: Duke University Press, 2019); Nikhil Anand, Akhil Gupta, and Hannah Appel, eds., *The Promise of Infrastructure* (Durham: Duke University Press, 2018); Lisa Parks and Nicole Starosielski, eds., *Signal Traffic: Critical Studies of Media Infrastructures* (Urbana: University of Illinois Press, 2015); Angela Mitropoulos, *Contract and Contagion: From Biopolitics to Oikonomia* (New York: Minor Compositions, 2012); Jean-Paul Addie, "The Times of Splintering Urbanism," *Journal of Urban Technology* 29, no. 1 (2022): 109–16; Lisa Björkman, *Pipe Politics, Contested Waters: Embedded Infrastructures of Millenial Mumbai* (Durham: Duke University Press, 2015); Roger Keil, ed., *Suburban Constellations: Governance, Land, and Infrastructure in the 21st Century* (Berlin: Jovis, 2013).

83. Langdon Winner, "Do Artifacts Have Politics?," *Daedalus* 109, no. 1 (1980): 121–36; Bruno Latour, *The Pasteurization of France* (Cambridge, Mass.: Harvard University Press, 1988); Stephen Graham and Colin McFarlane, eds., *Infrastructural Lives: Urban Infrastructure in Context* (Abingdon: Routledge, 2015); Braun, "A New Urban Dispositif?"; Michael Dillon and Julian Reid, *The Liberal Way of War: Killing to Make Life Live* (London: Routledge, 2009).

84. Michel Foucault, *The History of Sexuality: Volume 1* (New York: Vintage, 1990); Michel Foucault, *The Birth of Biopolitics: Lectures at the Collège de France, 1978–1979* (New York: Palgrave Macmillan, 2008); Foucault, "Confession of the Flesh"; Aggregate, *Governing by Design: Architecture, Economy, and Politics in the Twentieth Century* (Pittsburgh: University of Pittsburgh Press, 2012); Andrew Barry, *Political Machines: Governing a Technological Society* (London: Athlone, 2001); Patrick Joyce, *The Rule of Freedom: The City and Modern Liberalism* (London: Verso, 2003).

85. Larkin, "Politics and Poetics of Infrastructure," 328.

86. Easterling, *Extrastatecraft*, 14.

87. Yannis Kallianos, "Infrastructural Disorder: The Politics of Disruption, Contingency, and Normalcy in Waste Infrastructures in Athens," *Environment and Planning D: Society and Space* 36, no. 4 (2018): 3.

88. Swati Chattopadhyay, *Unlearning the City: Infrastructure in a New Optical Field* (Minneapolis: University of Minnesota Press, 2012), x.

89. Christine Boyer, *Dreaming the Rational City: The Myth of American City Planning* (Cambridge: MIT Press, 1986); Timothy Mitchell, *Colonizing Egypt* (Berkeley: University of California Press, 1991); Thomas Osborne, "Security and Vitality: Drains, Liberalism, and Power in the Nineteenth Century," in *Foucault and Political Reason: Liberalism, Neo-liberalism, and Rationalities of Government*, ed. Andrew Barry, Thomas Osborne, and Nikolas Rose (Chicago: University of Chicago Press, 1996), 99–121; Chris Otter, "Making Liberalism Durable: Vision and Civility in the Late Victorian City," *Social History* 27, no. 1 (2002): 1–15; Joyce, *The Rule of Freedom*.

90. Rudolf Mrázek, *Engineers of Happy Land: Technology and Nationalism in a Colony* (Princeton: Princeton University Press, 2002); Timothy Mitchell, *Carbon Democracy: Political Power in the Age of Oil* (London: Verso, 2011).

91. Larkin, "Politics and Poetics of Infrastructure," 332; Daniel Headrick, *The Tentacles of Progress: Technology Transfer in the Age of Imperialism, 1850–1940* (New York: Oxford University Press, 1988), 12.

92. Connolly, *A World More Concrete*, 136–37.

93. Stephen Collier and Andrew Lakoff, "The Vulnerability of Vital Systems: How 'Critical Infrastructure' Became a Security Problem," in *Securing the Homeland: Critical Infrastructure, Risk, and (In)security*, ed. Miriam Anna Dunn and Kristian Søby Kristensen (London: Routledge, 2008), 17–39; Graham and Marvin, *Splintering Urbanism;* Tony H. Grubesic and Alan T. Murray, "Vital Nodes, Interconnected Infrastructures, and the Geographies of Network Survivability," *Annals of the Association of American Geographers* 96, no. 1 (2006): 64–83; Jon Coaffee and Peter Lee, *Urban Resilience: Planning for Risk Crisis and Uncertainty* (London: Palgrave Macmillan, 2016), among others.

94. Chiara Bottici, *Imaginal Politics: Images beyond Imagination and the Imaginary* (New York: Columbia University Press, 2014), 4.

95. Julian Reid, "Securing the Imagination: The Politics of the Resilient Self," in *The Resilience Machine*, ed. Jim Bohland, Simin Davoudi, and Jennifer Lawrence (Abingdon: Routledge, 2018), 29

96. Charles Taylor, *Modern Social Imaginaries* (Durham: Duke University Press, 2004), 23; Edward Said, *Orientalism* (New York: Pantheon Books, 1978); Cornelius Castoriadis, *The Imaginary Institution of Society* (Cambridge: MIT Press, 1987).

97. See, e.g., David Harvey, *Social Justice and the City* (Baltimore:

John Hopkins University Press, 1973); Derek Gregory, *Geographical Imaginations* (Malden, Mass.: Wiley-Blackwell, 1994); Derek Gregory, "Geographical Imagination," in *The Dictionary of Human Geography*, ed. Derek Gregory, Ron Johnston, Geraldine Pratt, Michael Watts, and Sarah Whatmore (Chichester, UK: Wiley-Blackwell, 2009), 282–85; Jen Jack Gieseking, "Geographical Imagination," in *International Encyclopedia of Geography*, ed. Douglas Richardson, Noel Castree, Michael F. Goodchild, Audrey Kobayashi, Weidong Liu, and Richard A. Marston (New York: Wiley-Blackwell and the Association of American Geographers, 2017); Wendy Wolford, "This Land Is Ours Now: Spatial Imaginaries and the Struggle for Land in Brazil," *Annals of the American Association of Geographers* 94, no. 2 (2004): 409–24; Simin Davoudi, "Imagination and Spatial Imaginaries: A Conceptual Framework," in Simin Davoudi, Ruth Raynor, Bryonie Reid, Jenny Crawford, Olivier Sykes, and David Shaw, "Policy and Practice Spatial Imaginaries: Tyrannies or Transformations?," *Town Planning Review* 89, no. 2 (2018): 101.

98. Davoudi, "Imagination and Spatial Imaginaries," 101.

99. Stephen Daniels, "Geographical Imagination," *Transactions of the Institute of British Geographers* 36, no. 2 (2011): 182–87; Gieseking, "Geographical Imagination."

100. Craig Browne and Paula Diehl, "Conceptualising the Political Imaginary: An Introduction to the Special Issue," *Social Epistemology* 33, no. 5 (2019): 1–5; Denis E. Cosgrove and Veronica della Dora, "Mapping Global War: Los Angeles, the Pacific, and Charles Owens's Pictorial Cartography," *Annals of the Association of American Geographers* 95, no. 2 (2005): 373–90; Reid, "Securing the Imagination"; Sarah Whatmore, *Hybrid Geographies: Natures Cultures Spaces* (London: SAGE, 2002).

101. Daniels, "Geographical Imagination," 183; Kathryn Yusoff and Jennifer Gabrys, "Climate Change and the Imagination," *Wiley Interdisciplinary Reviews: Climate Change* 2, no. 4 (2011): 516–34.

102. Davoudi, "Imagination and Spatial Imaginaries."

103. Taylor Shelton, "The Urban Geographical Imagination in the Age of Big Data," *Big Data & Society* 4, no. 1 (2017): 4.

104. Shelton, 4–5.

105. Shelton, 8.

106. Davoudi, "Imagination and Spatial Imaginaries," 98.

107. See, e.g., Kevin Lynch, *The Image of the City* (Cambridge: MIT Press, 1960); Christoph Lindner and Miriam Meissner, eds., *The Routledge Companion to Urban Imaginaries* (London: Routledge Taylor &

Francis Group, 2019); Edward Soja, *Postmetropolis: Critical Studies of Cities and Regions* (Malden, Mass.: Blackwell, 2000); Nicholas A. Phelps, *The Urban Planning Imagination: A Critical International Introduction* (Cambridge, UK: Polity Press, 2021).

108. See, e.g., Daniels, "Geographical Imagination"; Gregory, *Geographical Imaginations;* Said, *Orientalism;* Edward Said, *Culture and Imperialism* (New York: Vintage Books, 1994); Martin Mahony and Samuel Randalls, eds., *Weather, Climate, and the Geographical Imagination: Placing Atmospheric Knowledges* (Pittsburgh: University of Pittsburgh Press, 2020); Harvey, *Social Justice;* Sheila Jasanoff and Sang-Hyun Kim, eds., *Dreamscapes of Modernity: Sociotechnical Imaginaries and the Fabrication of Power* (Chicago: University of Chicago Press, 2015); and Doreen Massey, *World City* (Cambridge: Polity Press, 2007); among others.

109. Mahony and Randalls, *Weather, Climate, and the Geographical Imagination.*

110. Earl Harper and Doug Specht, eds., *Imagining Apocalyptic Politics in the Anthropocene* (Abingdon: Routledge, 2021).

111. Erik Swyngedouw, "Apocalypse Forever? Post-political Populism and the Spectre of Climate Change," *Theory Culture Society* 27, nos. 2–3 (2010): 218.

112. Erik Swyngedouw, "Apocalypse Now! Fear and Doomsday Pleasures," *Capitalism Nature Socialism* 24, no. 1 (2013): 9–18; Swyngedouw, "Apocalypse Forever?"

113. Benjamin Bratton, *The Terraforming* (n.p.: Strelka Press, 2019).

114. Kasia Paprocki, "The Climate Change of Your Desires: Climate Migration and Imaginaries of Urban and Rural Climate Futures," *Environment and Planning D: Society and Space* 38, no. 2 (2020): 248–66.

115. Shelton, "The Urban Geographical Imagination."

116. Yusoff and Gabrys, "Climate Change and the Imagination." On Anthropocene imaginaries of "hope," see also Cameron Harrington, "The Eternal Return: Imagining Security Futures at the Doomsday Vault," *Environment and Planning E: Nature and Space* 6, no. 4 (2023): 2614–35.

117. David L. Levy and André Spicer, "Contested Imaginaries and the Cultural Political Economy of Climate Change," *Organization* 20 (2013): 662. See also Rebecca Lave, "The Future of Environmental Expertise," *Annals of the Association of American Geographers* 105, no. 2 (2015): 244–52; Ruth Fincher, Jon Barnett, and Sonia Graham, "Tem-

poralities in Adaptation to Sea-Level Rise," *Annals of the Association of American Geographers* 105, no. 2 (2015): 263–73.

118. Said, *Culture and Imperialism*, 7.

119. Hillary Angelo, "From the City Lens toward Urbanisation as a Way of Seeing: Country/City Binaries on an Urbanising Planet," *Urban Studies* 54, no. 1 (2017): 158–78; Hillary Angelo, *How Green Became Good: Urbanized Nature and the Making of Cities and Citizens* (Chicago: University of Chicago Press, 2021).

120. T. D. Allman, *Miami: City of the Future* (Gainesville: University Press of Florida, 1987), 105.

121. Joan Didion, *Miami* (New York: Simon and Schuster, 1987), 52.

122. Quoted in Jean-François Lejeune, "City without Memory: Planning the Spectacle of Greater Miami," in *Miami Modern Metropolis: Paradise and Paradox in Midcentury Architecture and Planning*, ed. Alan T. Shulman (Miami Beach: Bass Museum of Art and Glendale, Calif.: Balcony Press, 2009), 35.

123. Larkin, "Politics and Poetics of Infrastructure."

124. Laura Ogden, "The Everglades Ecosystem and the Politics of Nature," *American Anthropologist* 110, no. 1 (2008): 22; Laura Ogden, *Swamplife: People, Gators, and Mangroves Entangled in the Everglades* (Minneapolis: University of Minnesota Press, 2011); Jessica Cattelino, "The Cultural Politics of Water in the Everglades and Beyond," *Hau: Journal of Ethnographic Theory* 5, no. 3 (2015): 235–50.

125. Michael Truscello, *Infrastructural Brutalism: Art and the Necropolitics of Infrastructure* (Cambridge: MIT Press, 2020).

126. Anderson, "What Kind of Thing Is Resilience?," 62.

127. Rickards et al., "Urban Studies," 1524.

128. John Law, "What's Wrong with a One-World World?," *Distinktion: Journal of Social Theory* 16, no. 1 (2015): 126–39.

129. Christopher Flavelle and Patricia Mazzei, "Miami Says It Can Adapt to Rising Seas. Not Everyone Is Convinced," *New York Times*, March 2, 2021, https://www.nytimes.com/2021/03/02/climate/miami-sea-level-rise.html.

1. Miami Imaginaries

1. David McCally, *Everglades: An Environmental History* (Gainesville: University Press of Florida, 1999).

2. Gregory Bush, "'Playground of the USA': Miami and the Promotion of Spectacle," *Pacific Historical Review* 68, no. 2 (1999): 156;

Jan Nijman, *Miami: Mistress of the Americas* (Philadelphia: University of Pennsylvania Press, 2011), 23.

3. N. D. B. Connolly, *A World More Concrete: Real Estate and the Remaking of Jim Crow South Florida* (Chicago: University of Chicago Press, 2014).

4. Alex Harris, "Octopus in the Parking Garage Is Climate Change's Canary in the Coal Mine," *Miami Herald*, November 22, 2019, https://www.miamiherald.com/news/local/community/miami-dade/miami-beach/article115688508.html#storylink=cpy.

5. See, e.g., Kevin Loria, "Miami Is Racing against Time to Keep Up with Sea-Level Rise," *Business Insider*, April 12, 2018, https://www.businessinsider.com/miami-floods-sea-level-rise-solutions-2018-4.

6. See, e.g., "Sea Level Rise (SLR) Building Impacts," Miami-Dade County website, https://mdc.maps.arcgis.com/apps/webappviewer3d/index.html?id=b92a9fa4ff8847bf97f3e628a195a398.

7. Peter Harlem, "Sea Level Rise Map Projections," *Sea Level Rise Collection* 2, 2008, https://digitalcommons.fiu.edu/sea_level_rise/2.

8. Robin McKie, "Miami, the Great World City, Is Drowning while the Powers That Be Look Away," *Guardian*, July 11, 2014, https://www.theguardian.com/world/2014/jul/11/miami-drowning-climate-change-deniers-sea-levels-rising.

9. Harold Wanless, "Statement on Sea Level in the Coming Century," Science Committee Miami-Dade County Climate Change Advisory Task Force to the Board of County Commissioners, April 22, 2008, https://www.miamidade.gov/environment/library/brochures/08-04-22-statement-on-sea-level.pdf; see also Harold Wanless, "The Coming Reality of Sea Level Rise: Too Fast Too Soon," November 2014, white paper, https://arboretum.as.miami.edu/_assets/pdf/wanless.pdf; Ariel Schwartz, "Millennials Are Flocking to Miami, Where They Might All Drown Soon," *Fast Company*, November 12, 2014, https://www.fastcompany.com/3037978/millennials-are-flocking-to-miami-where-they-might-all-drown-soon; Juliet Pinto and Kate MacMillin, "[Documentary] South Florida's Rising Seas," *Sea Level Rise Collection* 76, 2014, https://digitalcommons.fiu.edu/sea_level_rise/76.

10. Southeast Florida Regional Climate Change Compact Counties Technical Ad Hoc Work Group, "A Unified Sea Level Rise Projection for Southeast Florida," document prepared for the Southeast Florida Regional Climate Change Compact Steering Committee," April 2011, https://southeastfloridaclimatecompact.org/wp-content/uploads/2023/10/2011-sea-level-projections.pdf.

NOTES TO CHAPTER 1 | 225

11. Southeast Florida Regional Climate Change Compact Sea Level Rise Ad Hoc Work Group, *Unified Sea Level Rise Projection for Southeast Florida*, 2019 update, 10, https://southeastfloridaclimatecompact.org/wp-content/uploads/2020/04/Sea-Level-Rise-Projection-Guidance-Report_FINAL_02212020.pdf.

12. Jeff Goodell, "Miami: How Rising Sea Levels Endanger South Florida," *Rolling Stone*, June 2013, https://www.rollingstone.com/feature/miami-how-rising-sea-levels-endanger-south-florida-200956/.

13. Goodell.

14. See, e.g., Lars Jan, *Holoscenes*, 2015, performance art installation, MDC Live Arts at Miami-Dade College Wolfson Campus; Xavier Cortada, *Underwater HOA*, 2018, public art exhibition, Village of Pinecrest, Miami, Florida; Leandro Erlich, *Order of Importance*, 2019, public art installation commissioned by City of Miami Beach, Miami Beach, Florida.

15. Goodell, "Miami."

16. Miami-Dade County, *Sea Level Rise Strategy*, Miami-Dade County Office of Resilience, 2021, https://miami-dade-county-sea-level-rise-strategy-draft-mdc.hub.arcgis.com/.

17. Miami-Dade County, *Miami-Dade County Sea Level Rise Strategy*, Miami-Dade County Office of Resilience, 2021, 30–49, https://mdc.maps.arcgis.com/sharing/rest/content/items/b1a93c86abb548f99f712df2dac6d670/data.

18. Stan Cox and Paul Cox, *How the World Breaks: Life in Catastrophe's Path, from the Caribbean to Siberia* (New York: The New Press, 2016); Kate Stein, "'We're a Living Laboratory': Miami Beach Works on Resiliency as Businesses Face Flooding," *WLRN*, October 17, 2016, https://www.wlrn.org/news/2016-10-17/were-a-living-laboratory-miami-beach-works-on-resiliency-as-businesses-face-flooding; Greater Miami and the Beaches, *Resilient 305*, May 2019, https://resilient305.com/; 100 Resilient Cities, Columbia University's Center for Resilient Cities and Landscapes, and Greater Miami and the Beaches, *Resilient Accelerator Southeast Florida: Workshop Report, Designing for Regional Resilience*, August 2018, https://crcl.columbia.edu/sites/default/files/content/Miami/Workshop%20Photos/Resilience%20Accelerator%20Workshop%20Report_Southeast%20Florida_WEB-compressed.pdf; Aspen Institute, "Aspen Ideas: Climate," 2023, https://www.aspenideas.org/pages/aspen-ideas-climate.

19. City of Miami Beach, "Miami Beach Rising Above," 2022, video, https://www.mbrisingabove.com/your-city-at-work/resilience-strategy/;

Union of Concerned Scientists and Rethink Energy Florida, "Florida: Ground Zero in the Climate Crisis," 4, https://www.ucsusa.org/sites/default/files/attach/2019/05/Florida-Gound-Zero-in-the-Climate-Crisis-newer.pdf; Urban Land Institute, *Stormwater Management and Climate Adaptation Review: An Urban Land Institute Advisory Services Panel Report*, Miami Beach, FL, April 16–19; "Philip Levine for Mayor of Miami Beach 'Paddle,'" campaign ad, 2013, https://www.youtube.com/watch?v=N9niAnh9KZw.

20. David Smiley and Joey Flechas, "When It Rains in Miami, the Politicians Pile It On," *Miami Herald*, August 2, 2017, https://www.miamiherald.com/news/local/community/miami-dade/article165104327.html; City of Miami Beach, "Miami Beach Rising Above."

21. City of Miami Beach, "Rising above the Risk FAQ," Report, May 5, 2017, 1, https://www.miamibeachfl.gov/wp-content/uploads/2017/08/Rising-Above-the-Risk_FAQs-05052017.pdf.

22. Jacobs, "Miami Beach Integrated Water Management Neighborhood Project Prioritization—Methodology and Results," Final Report, April 1, 2020, City of Miami Beach, RFQ 2018–312-KB, https://www.miamibeachfl.gov/wp-content/uploads/2020/07/Neighborhood-Project-Prioritization.pdf; Alex Harris, "Raising Flood-Prone Roads Has Angered Miami Beach Residents: Experts Say They Need to Go Higher," *Miami Herald*, January 22, 2022, https://www.miamiherald.com/news/local/environment/article239486308.html#storylink=cpy.

23. City of Miami Beach, "Rising above the Risk FAQ."

24. Alex Harris, "Miami Beach Wants Higher Roads and Pumps to Fight Sea Rise. Some Residents Say No Way," *Miami Herald*, May 16, 2018, https://www.miamiherald.com/news/local/community/miami-dade/miami-beach/article211237324.

25. Joey Flechas and Jenny Staletovich, "Miami's Battle to Stem Rising Tides," *Miami Herald*, October 23, 2015, https://www.miamiherald.com/news/local/community/miami-dade/miami-beach/article41141856.html; Harris, "Miami Beach Wants Higher Roads."

26. Joshua Evans, "Trials and Tribulations: Conceptualizing the City through/as Urban Experimentation," *Geography Compass* 10, no. 10 (2016): 429–43; Harriet Bulkeley, Vanesa Castán Broto, and Gareth Edwards, *An Urban Politics of Climate Change: Experimentation and the Governing of Socio-Technical Transitions* (Abingdon: Routledge, 2015); Vanesa Castán Broto and Harriet Bulkeley, "A Survey of Urban Climate Change Experiments in 100 Cities," *Global Environmental*

Change 23, no. 1 (2013): 92–102; James Evans, Andrew Karvonen, and Rob Raven, *The Experimental City* (Abingdon: Routledge, 2016); James Evans and Andrew Karvonen, "'Give me a laboratory and I will lower your carbon footprint!'—Urban Laboratories and the Governance of Low-Carbon Futures," *International Journal of Urban and Regional Resilience* 38, no. 2 (2013): 413–30; Andrew Karvonen, James Evans, and Bas van Heur, "The Politics of Urban Experiments: Radical Change or Business as Usual?," in *After Sustainable Cities*, ed. Simon Marvin and Mike Hodson, (Abingdon: Routledge, 2014), 105–14.

27. See, e.g., Thomas F. Gieryn, "City as Truth-Spot: Laboratories and Field-Sites in Urban Studies," *Social Studies of Science* 36, no. 1 (2006): 5–38.

28. Angela Oels, "Rendering Climate Change Governable by Risk: From Probability to Contingency," *Geoforum* 45 (2013): 17–29; Matthew J. Hoffmann, *Climate Governance at the Crossroads: Experimenting with a Global Response after Kyoto* (Oxford: Oxford University Press, 2011); Jonas Torrens and Timo von Wirth, "Experimentation or Projectification of Urban Change? A Critical Appraisal and Three Steps Forward," *Urban Transformations* 3, no. 8 (2021), https://doi.org/10.1186/s42854-021-00025-1.

29. Harriet Bulkeley and Vanesa Castán Broto, "Government by Experiment? Global Cities and the Governing of Climate Change," *Transactions of the Institute of British Geographers* 38 (2013): 361–75; James Evans, "Resilience, Ecology, and Adaptation in the Experimental City," *Transactions of the Institute of British Geographers* 36 (2011): 223–37.

30. James Evans, "Resilience, Ecology, and Adaptation."

31. Simon Dalby, *Anthropocene Geopolitics: Globalization, Security, Sustainability* (Ottawa: University of Ottawa Press, 2020).

32. Simon Dalby, "Biopolitics and Climate Security in the Anthropocene," *Geoforum* 49 (2013): 189.

33. On experimentation as urban knowledge innovation, see Evans and Karvonen, "'Give me a laboratory'"; "provisional, risky, and dynamic" from Karvonen, Evans, and van Heur, "Politics of Urban Experiments," 2–3; "explore uncharted policy territories" from Vanesa Castán Broto and Harriet Bulkeley, "Maintaining Climate Change Experiments: Urban Political Ecology and the Everyday Reconfiguration of Urban Infrastructure," *International Journal of Urban and Regional Research* 37, no. 6 (2013): 1935.

34. Evans and Karvonen, "'Give me a laboratory'"; James Evans, "Resilience, Ecology, and Adaptation"; Broto and Bulkeley, "Maintaining Climat Change Experiments."

35. Broto and Bulkeley, "Survey of Urban Climate Change Experiments."

36. James Evans, "Resilience, Ecology, and Adaptation"; Joshua Evans, "Trials and Tribulations."

37. Karvonen, Evans, and van Heur, "Politics of Urban Experiments," 5.

38. Stephanie Wakefield and Bruce Braun, "Oystertecture: Infrastructure, Profanation, and the Sacred Figure of the Human," in *Infrastructure, Environment, and Life in the Anthropocene*, ed. Kregg Hetherington (Durham: Duke University Press, 2019), 193–215; Joshua Evans, "Trials and Tribulations," 437; Broto and Bulkeley, "Survey of Urban Climate Change Experiments"; Timon McPhearson, Nadja Kabisch, and Niki Frantzeskaki, eds., *Nature-Based Solutions for Cities* (Northampton, Mass.: Edward Elgar Publishing, 2023).

39. Alex Harris, "Miami Beach's Future Is 'Uncertain,' Experts Say, but Sea Rise Pumps Are a Good Start," *Miami Herald*, April 19, 2018, https://www.miamiherald.com/news/local/community/miami-dade/miami-beach/article209328849.html.

40. Urban Land Institute, *Stormwater Management*, 15.

41. David Smiley, Alex Harris, and Joey Flechas, "Levine Is Campaigning on Climate Change, but His Sea-Rise Legacy Is Complicated," *Miami Herald*, May 24, 2018, https://www.miamiherald.com/article211326029.html.

42. Gareth Edwards and Harriet Bulkeley, "Heterotopia and the Urban Politics of Climate Change Experimentation," *Environment and Planning D: Society and Space* 36, no. 2 (2018): 356; Evans and Karvonen, "'Give me a laboratory.'"

43. Karvonen, Evans, and van Heur, "Politics of Urban Experiments," 8; Raymond De Young and Stephen Kaplan, "Adaptive Muddling," in *The Localization Reader: Adapting to the Coming Downshift*, ed. Raymond De Young and Thomas Princen (Cambridge: MIT Press, 2012), 287–98.

44. Amanda Ruggeri, "Miami's Fight against Rising Seas," *BBC Future Now*, May 22, 2020, https://www.bbc.com/future/article/20170403-miamis-fight-against-sea-level-rise; "An Idea to Mitigate Rising Seas in Miami Beach: Lift the Entire City," *WLRN*, March 20, 2015, http://

www.wlrn.org/post/idea-mitigate-rising-seas-miami-beach-lift-entire-city.

45. Smiley, Harris, and Flechas, "Levine Is Campaigning."
46. Flechas and Staletovich, "Miami's Battle."
47. Urban Land Institute, *Stormwater Management*.
48. Urban Land Institute, 14 ("embracing"), 18 ("tourism and economic").
49. Andres Viglucci, "He Kept the Netherlands Dry. Now He Aims to Defend Miami and the World from Rising Seas," *Miami Herald*, February 4, 2017, https://www.miamiherald.com/news/local/community/miami-dade/article132057654.html.
50. The YEARS Project, "Jack Black Meets Florida Mayors," video excerpt from *Years of Living Dangerously*, May 16, 2018, https://www.youtube.com/watch?v=T-RxBV83X1o.
51. Smiley, Harris, and Flechas, "Levine Is Campaigning."
52. Flechas and Staletovich, "Miami's Battle."
53. Harriet Bulkeley, Simon Marvin, Yuliya Voytenko Palgan, Kes McCormick, Marija Breitfuss-Loidl, Lindsay Mai, Timo Von Wirth, and Niki Frantzeskaki, "Urban Living Laboratories: Conducting the Experimental City?," *European Urban and Regional Studies* 26, no. 4 (2019): 317–35.
54. Stephanie Wakefield, "Urban Resilience as Critique: Problematizing Infrastructure in Post-Sandy New York City," *Political Geography* 79 (2020): 102148; Ross Exo Adams, "Notes from the Resilient City," *Log*, no. 32 (2014); Andrew Zolli, "Learning to Bounce Back," *New York Times*, November 2, 2012, http://www.nytimes.com/2012/11/03/opinion/forget-sustainability-its-about-resilience.html.
55. Smiley, Harris, and Flechas, "Levine Is Campaigning."
56. Harris, "Raising Flood-Prone Roads."
57. Connolly, *A World More Concrete*; Harvey Molotch, "The City as a Growth Machine: Toward a Political Economy of Place," *American Journal of Sociology* 82, no. 2 (1976): 309–32; Jan Nijman, "Globalization to a Latin Beat: The Miami Growth Machine," *Annals of the American Academy of Political and Social Science* 551, no. 1 (1997): 164–77; Saskia Sassen and Alejandro Portes, "Miami: A New Global City?," *Contemporary Sociology* 22, no. 4 (1993): 471–77; Lars Ankum, Wessel Brocken, and Tiemen Koch, "'Don't Count on Living in Miami': A Human Geographical Approach to Understanding the Ramifications of Sea Level Rise for Urban Miami," *Sea Level Rise Collection* 52,

2016, http://digitalcommons.fiu.edu/sea_level_rise/52; Jason Vuic, *The Swamp Peddlers: How Lot Sellers, Land Scammers, and Retirees Built Modern Florida and Transformed the American Dream* (Chapel Hill: University of North Carolina Press, 2021), 16; Zac Taylor and Manuel Aalbers, "Climate Gentrification: Risk, Rent, and Restructuring in Greater Miami," *Annals of the American Association of Geographers* 112, no. 6 (2022): 1685–1701.

58. Alejandro Portes and Ariel C. Armony, *The Global Edge: Miami in the 21st Century* (Oakland: University of California Press, 2018), 73; Nijman, "Globalization to a Latin Beat"; Nicholas Nehamas, "How Secret Offshore Money Helps Fuel Miami's Luxury Real-Estate Boom," *Miami Herald*, April 3, 2016, https://www.miamiherald.com/news/business/real-estate-news/article69248462.html.

59. David Madden, 'The Urban Process under COVID Capitalism," *City* 24, nos. 5–6 (2020): 677.

60. Ayona Datta, "Self(ie)-Governance: Technologies of Intimate Surveillance in India under COVID-19," *Dialogues in Human Geography* 10, no. 2 (2020): 234–37; Bei Chen, Simon Marvin, and Aidan While, "Containing COVID-19 in China: AI and the Robotic Restructuring of Future Cities," *Dialogues in Human Geography* 10, no. 2 (2020): 238–41; Ola Söderström, "The Three Modes of Existence of the Pandemic Smart City," *Urban Geography* 42, no. 3 (2021): 399–407.

61. Tom Carter, "Billionaire Ken Griffin Says Miami Could Overtake New York as a Financial Hub," *Business Insider*, November 15, 2023, https://www.businessinsider.com/billionaire-ken-griffin-miami-will-overtake-new-york-finance-hub-2023-11#:~:text=The%20area%20has%20been%20nicknamed,to%20sing%20Miami's%20praises%20recently.

62. Connie Ogle, "Miami Is the Most Glamorous City in America, This Study Says. We Choose to Believe It," *Miami Herald*, October 13, 2021, https://www.miamiherald.com/miami-com/miami-com-news/article248771740.html.

63. Stacey Lastoe, "Miami: So Hot Right Now," *Time*, July 12, 2022, https://time.com/collection/worlds-greatest-places-2022/6194452/miami/.

64. Joel Stein, "How Miami Became the Most Important City in America," *Financial Times*, https://www.ft.com/content/77ee0d8d-bf74-4cc3-bde0-a064ce074726.

65. Taylor and Aalbers, "Climate Gentrification"; Jesse Keenan, Thomas Hill, and Anurag Gumber, "Climate Gentrification: From The-

ory to Empiricism in Miami-Dade County, Florida," *Environmental Research Letters* 13, no. 5 (2018): 054001.

66. Benjamin Wallace, "How Miami Seduced Silicon Valley," *New York Magazine*, September 21, 2021, https://nymag.com/intelligencer/2021/09/how-miami-seduced-silicon-valley.html.

67. Francis Suarez, Twitter post, April 6, 2022, 12:32 p.m., https://twitter.com/FrancisSuarez/status/1511743745837998083?s=20&t=Q-5tsavfk7BR18y4FcH2aw.

68. Joey Flechas and Vinod Sreeharsha, "MiamiCoin Trading Halted. After Price Tanked, Mayor Francis Suarez Lost about $2,500," *Miami Herald*, April 23, 2023, https://www.miamiherald.com/news/local/community/miami-dade/article273473595.html#storylink=cpy.

69. Rabois details from Wallace, "How Miami Seduced Silicon Valley."

70. Bill Kearney, "Miami's Little Manhattan," *Curbed*, March 16, 2021: https://www.curbed.com/2021/03/wealthy-new-yorkers-moving-to-south-florida.html; Jennifer Sangalang, "'Billionaire Bunker' Florida Home Listed at $85 million. Jeff Bezos got it for $79 Million," *Palm Beach Post*, October 24, 2023, https://www.palmbeachpost.com/story/news/2023/10/24/jeff-bezos-billionaire-bunker-florida-indian-creek-village-amazon-blue-origin-tom-brady-ivanka/71238228007/.

71. Svea Herbst-Bayliss, "Hedge Fund Citadel to Move Headquarters to Miami from Chicago," *Reuters*, June 23, 2022, https://www.reuters.com/business/hedge-fund-citadel-move-headquarters-miami-chicago-2022-06-23/.

72. RealtyHop, "RealtyHop Housing Affordability Index: February 2022," RealtyHop blog, March 4, 2022, https://www.realtyhop.com/blog/housing-affordability-index-february-2022/.

73. State of Florida Executive Office of Governor Ron DeSantis, "Florida Breaks Record for Highest Visitation in State History," press release, November 17, 2022, https://www.flgov.com/2022/11/17/florida-tourism-continues-to-experience-record-growth-in-q3-2022/.

74. This number includes both overnight and day visitors. Greater Miami Convention & Visitors Bureau, *Greater Miami and Miami Beach: 2021 Visitor Industry Overview*, 5, https://www.miamiandbeaches.com/getmedia/fcf55ce5-3058-4362-990b-e33cf85171d2/Visitor_Industry_Overview_2021_FINAL_1.pdf.aspx.

75. Greater Miami Convention & Visitors Bureau, 68.

76. Greater Miami and the Beaches, "Greater Miami Convention

& Visitors Bureau (Gmcvb) Celebrates Unmatched Visitor Growth and Economic Impact in 2022," press release, May 11, 2023, https://www.miamiandbeaches.com/press-and-media/miami-press-releases/growth-and-economic-impact.

77. Greater Miami Convention & Visitors Bureau, *Greater Miami*, 78.

78. Risky Business Project, *The Economic Risks of Climate Change in the United States*, June 2014, 24, https://riskybusiness.org/site/assets/uploads/2015/09/RiskyBusiness_Report_WEB_09_08_14.pdf.

79. Sam Paustian-Underdahl, Carlos Parra, Ronald Mesia, Aris Papadopoulos, Nancy Scanlon, and Tiffany Troxler, *Getting Ahead of the Curve for a Resilient Economy: An Introduction to Risks and Opportunities of Climate Change for Top Business Sectors in Southeast Florida*, Sea Level Solutions Center Publication no. 8, Institute of Water and Environment, Florida International University (updated 2018), 28, https://environment.fiu.edu/slsc/resources/_assets/slsc.businessschool.report_final5.pdf.

80. Urban Land Institute, *The Business Case for Resilience in Southeast Florida: Executive Summary* (Washington, D.C.: Urban Land Institute, 2021), 3.

81. Urban Land Institute, 3.

82. Urban Land Institute, 8.

83. Urban Land Institute, 13. See also "Watch: Developers and Brokers Weigh In on Miami Sea-Level Rise," *The Real Deal*, October 23, 2019, https://therealdeal.com/miami/2019/10/23/watch-developers-and-brokers-weigh-in-on-miami-sea-level-rise/.

84. Andrew Morrell, "This Week in Miami Real Estate: Condo Lawsuits, Sea Level Rise and More," *South Florida Agent Magazine*, November 29, 2018, https://southfloridaagentmagazine.com/2018/11/29/week-miami-real-estate-condo-lawsuits-sea-level-rise/. On PAMM's climate adaptive architecture, see Monica Isola, "Museum Protects Art from Sea-Level Rise," *Yale Climate Connections*, October 5, 2018, https://yaleclimateconnections.org/2018/10/museum-protects-art-from-sea-level-rise/.

85. Amy Viteri, "Controversy Arising in Miami Beach over Raising Roads to Combat Flooding," WPLG Local10.com, February 21, 2022, https://www.local10.com/news/local/2022/02/21/controversy-arising-in-miami-beach-over-raising-roads-to-combat-flooding/.

86. Alex Harris, "Miami Wants to Make the Case to the World That Climate-Proofing Is Worth the Cost," *Miami Herald*, February 19, 2019, https://www.miamiherald.com/article226341990.html.

87. Jonathan Levin, "Dear Miami, Taking Wall Street from NYC

Won't Be Easy," *Bloomberg*, November 17, 2023, https://www.bloomberg.com/opinion/articles/2023-11-17/miami-is-a-novice-in-quest-to-build-wall-street-south.

88. Levin.

89. Savannah Cox, "Inscriptions of Resilience: Bond Ratings and the Government of Climate Risk in Greater Miami, Florida," *Environment and Planning A: Economy and Space* 54, no. 2 (2022): 6.

90. Paustian-Underdahl et al., *Getting Ahead of the Curve*.

91. Robert Meyer, "Miami and the Costs of Climate Change," University of Pennsylvania Wharton Climate Center blog, April 11, 2015, https://esg.wharton.upenn.edu/climate-center/miami-and-the-costs-of-climate-change/.

92. City of Miami Beach, "Miami Beach Rising Above."

93. In the Sunset Harbour neighborhood. Alex Harris, "Can Raising Roads for Sea Rise Make a Home More Valuable? Miami Beach Report Says Yes," *Miami Herald*, March 15, 2022, https://www.miamiherald.com/news/local/environment/article239682778.html.

94. City of Miami Beach, "G.O. Bond Credit Ratings," Report, April 2019, p. 1.

95. City of Miami Beach, 1.

96. Cox, "Inscriptions of Resilience," 2.

97. First Street Foundation, *The 9th National Risk Assessment: The Insurance Issue*, September 20, 2023, 30, https://report.firststreet.org/9th-National-Risk-Assessment-The-Insurance-Issue.pdf.

98. Khristopher J. Brooks, "AAA Pulls Back from Renewing Some Insurance Policies in Florida," *CBS News*, July 18, 2023, https://www.cbsnews.com/news/aaa-insurance-policies-florida-nonrenewal/.

99. First Street Foundation, *The 9th National Risk Assessment*, 20.

100. Alex Harris and Nicolás Rivero, "New Study Warns of 'Climate Insurance Bubble': Is That Driving Costs Up in Florida?," *Miami Herald*, September 20, 2023, https://www.miamiherald.com/news/local/environment/climate-change/article279490394.html.

101. Smiley, Harris, and Fletchas, "Levine Is Campaigning."

102. City of Miami Beach, Office of Inspector Geeral, Final Report, OIG Report no. 21-15, Indian Creek Project, November 15, 2022, p. 52, https://www.miamibeachfl.gov/wp-content/uploads/2020/04/OIG-No.-21-15-Indian-Creek-Project.pdf.

103. Lance Gunderson and C. S. Holling, eds., *Panarchy: Understanding Transformations in Systems of Humans and Nature* (Washington D.C.: Island Press, 2002).

104. Lance Gunderson, C. S. Holling, and Stephen Light, *Barriers and Bridges to the Renewal of Ecosystems and Institutions* (New York: Columbia University Press, 1995).

105. Sandi Zellmer and Lance Gunderson, "Why Resilience May Not Always Be a Good Thing: Lessons in Ecosystem Restoration from Glen Canyon and the Everglades," *Nebraska Law Review* 87, no. 4 (2009): 893.

106. Ben Anderson, "Preemption, Precaution, Preparedness: Anticipatory Action and Future Geographies," *Progress in Human Geography* 34, no. 6 (2010): 777–98; Peter Adey and Ben Anderson, "Anticipating Emergencies: Technologies of Preparedness and the Matter of Security," *Security Dialogue* 43, no. 2 (2012): 99–117.

107. Claudia Aradau and Rens Van Munster, "Governing Terrorism through Risk: Taking Precautions, (Un)knowing the Future," *European Journal of International Relations* 13, no. 1 (2007): 89–115; Ulrich Beck, "The Terrorist Threat: World Risk Society Revisited," *Theory, Culture, Society* 19, no. 4 (2002): 39–55.

108. Stephen Collier, "Enacting Catastrophe: Preparedness, Insurance, Budgetary Rationalization," *Economy and Society* 37, no. 2 (2008): 224–50; Mitchell Dean, *Governmentality: Power and Rule in Modern Society* (London, Sage, 1999); Nikolas Rose, "The Politics of Life Itself," *Theory, Culture, & Society* 28, no. 6 (2001): 1–30.

109. Aradau and Van Munster, "Governing Terrorism through Risk"; Collier, "Enacting Catastrophe."

110. Oels, "Rendering Climate Change Governable."

111. Kasia Paprocki, "All That Is Solid Melts into the Bay: Anticipatory Ruination and Climate Change Adaptation," *Antipode* 51, no. 1 (2019): 295–315.

112. Anderson, "Preemption, Precaution, Preparedness"; on "enactment," see Collier, "Enacting Catastrophe"; and Adey and Anderson, "Anticipating Emergencies."

113. Gareth Edwards and Harriet Bulkeley, "Heterotopia and the Urban Politics of Climate Change Experimentation," *Environment and Planning D: Society and Space* 36, no. 2 (2018): 356.

114. Anderson, "Preemption, Precaution, Preparedness"; Claudia Aradau and Rens Van Munster, *Politics of Catastrophe: Genealogies of the Unknown* (Abingdon: Routledge, 2011).

115. Edwards and Bulkeley, "Heterotopia."

116. "spaces of utopian hope" from Edwards and Bulkeley, "Hetero-

topia," 366; "averting looming ecological catastrophe" from Edwards and Bulkeley, "Heterotopia," 352

117. Evans and Karvonen, "'Give me a laboratory,'" 425.

118. Karvonen, Evans, and van Heur, "Politics of Urban Experiments."

119. Joshue Evans, "Trials and Tribulations," 435.

120. Andrew Karvonen and Bas van Heur, "Urban Laboratories: Experiments in Reworking Cities," *International Journal of Urban and Regional Research* 38, no. 2 (2014): 387.

121. Johan Rockström, Will Steffen, Kevin Noone, Åsa Persson, F. Stuart III Chapin, Eric Lambin, Timothy M. Lenton, Marten Scheffer, Carl Folke, Hans Joachim Schellnhuber, Björn Nykvist, Cynthia A. de Wit, Terry Hughes, Sander van der Leeuw, Henning Rodhe, Sverker Sörlin, Peter K. Snyder, Robert Costanza, Uno Svedin, Malin Falkenmark, Louise Karlberg, Robert W. Corell, Victoria J. Fabry, James Hansen, Brian Walker, Diana Liverman, Katherine Richardson, Paul Crutzen, and Jonathan Foley, "Planetary Boundaries: Exploring the Safe Operating Space for Humanity," *Ecology and Society* 14, no. 2 (2009): 32.

122. Jamie Lorimer, *Wildlife in the Anthropocene: Conservation after Nature* (Minneapolis: University of Minnesota Press, 2015); see also Michel Callon, Pierre Lascoumes, and Yannick Barthe, *Acting in an Uncertain World: An Essay on Technical Democracy* (Cambridge: MIT Press, 2009); Joshua Evans, "Trials and Tribulations."

123. Evans and Karvonen, "'Give me a Laboratory,'" 427.

124. James Evans, "Resilience, Ecology, and Adaptation."

125. Evans, 230.

126. Harris, "Miami Beach Wants Higher Roads."

127. Harris, "Miami Beach Wants Higher Roads."

128. Smiley, Harris, and Fletchas, "Levine Is Campaigning."

129. Quote from Smiley, Harris, and Fletchas; Toluse Olorunnipa, "Heat's Bosh Buys $12.5 Million House in Miami Beach," *South Florida Sun Sentinel*, September 24, 2010, https://www.sun-sentinel.com/business/fl-xpm-2010-09-24-fl-bosh-mansion-20100924-story.html.

130. Katherine Kallergis, "Crypto CEO Buying Chris Bosh's Former Waterfront Miami Beach Mansion Asking $42M," *The Real Deal*, December 28, 2021, https://therealdeal.com/miami/2021/12/28/crypto-ceo-buying-chris-boshs-former-waterfront-miami-beach-mansion-asking-42m/.

131. N. C. Millette, C. Kelble, A. Linhoss, et al., "Using Spatial Vari-

ability in the Rate of Change of Chlorophyll a to Improve Water Quality Management in a Subtropical Oligotrophic Estuary," *Estuaries and Coasts* 42 (2019): 1792–1803.

132. Jenny Staletovich, "Miami Beach King Tides Flush Human Waste into Bay, Study Finds," *Miami Herald*, May 16, 2016, https://www.miamiherald.com/news/local/environment/article77978817.html; Jenny Staletovich, "Miami Beach Attacks 'Sloppy Science' on Human Waste Bacteria in Bay," *Miami Herald*, June 8, 2016, https://www.miamiherald.com/news/local/environment/article82648642.html; Smiley, Harris, and Flechas, "Levine Is Campaigning."

133. Jerry Iannelli, "Why Did Miami Beach's Multimillion-Dollar Anti-flood Pumps Fail?," *Miami New Times*, August 2, 2017, https://www.miaminewtimes.com/news/miami-beach-floods-from-tropical-storm-emily-overwhelm-sea-level-rise-pumps-9543575.

134. Smiley, Harris, and Flechas, "Levine Is Campaigning"; Ina Cordle, "Fired Miami Beach Engineer Who Skirted Environmental Regulations Won't Lose His License," *The Real Deal*, April 5, 2018, https://therealdeal.com/miami/2018/04/05/fired-miami-beach-engineer-who-skirted-environmental-regulations-wont-lose-his-license/.

135. Smiley, Harris, and Flechas, "Levine Is Campaigning"; Cordle, "Fired"; Aaron Leibowitz and Alex Harris, "Pressure from Ex–Miami Beach Mayor a Factor in Sea-Rise Project's Problems, Report Says," *Miami Herald*, November 16, 2022, https://www.miamiherald.com/news/local/community/miami-dade/miami-beach/article268202532.html#storylink=cpy.

136. Adey and Anderson, "Anticipating Emergencies," 99.

137. James Brassett and Nick Vaughan-Williams, "Security and the Performative Politics of Resilience: Critical Infrastructure Protection and Humanitarian Emergency Preparedness," *Security Dialogue* 46, no. 1 (2015): 39.

138. Dalby, "Biopolitics and Climate Security," 184.

139. Edwards and Bulkeley, "Heterotopia."

140. Vuic, *The Swamp Peddlers*, 6.

141. Morton C. Paulson, *The Great Land Hustle* (Chicago: Henry Regnery, 1972), 46, quoted in Vuic, *The Swamp Peddlers*, 16.

142. Kathryn Yusoff and Jennifer Gabrys, "Climate Change and the Imagination," *Wiley Interdisciplinary Reviews: Climate Change* 2, no. 4 (2011): 518.

143. Per Olsson, Victor Galaz, and Wiebren Boonstra, "Sustainabil-

ity Transformations: A Resilience Perspective," *Ecology and Society* 19, no. 4 (2015), http://www.ecologyandsociety.org/vol19/iss4/art1/; David Iwaniec, Elizabeth M. Cook, Olga Barbosa, and Nancy B. Grimm, "The Framing of Urban Sustainability Transformations," *Sustainability* vol. 11, no. 3 (2019): 573, https://doi.org/10.3390/su11030573; Asad Asadzadeh, Amir Reza Khavarian-Garmsir, Ayyoob Sharifi, Pourya Salehi, and Theo Kötter, "Transformative Resilience: An Overview of Its Structure, Evolution, and Trends," *Sustainability* 14, no. 22 (2022): 15267, https://doi.org/10.3390/su142215267; Timon McPhearson, David M. Iwaniec, Zoé A. Hamstead, Marta Berbés-Blázquez, Elizabeth M. Cook, Tischa A. Muñoz-Erickson, Lelani Mannetti, and Nancy Grimm, "A Vision for Resilient Urban Futures," in *Resilient Urban Futures*, ed. Zoé A. Hamstead, David M. Iwaniec, Timon McPhearson, Marta Berbés-Blázquez, Elizabeth M. Cook, and Tischa A. Muñoz-Erickson (Cham, Switzerland: Springer International Publishing, 2021), 173–83.

2. New Houses for a New Century

1. George Sánchez-Calderón, *How to Win Friends and Influence People*, site-specific installation and performance at Faena Beach Ritual Burn, Art Basel, Miami, Florida, 2018, https://vimeo.com/380768482.

2. N. D. B. Connolly, *A World More Concrete: Real Estate and the Remaking of Jim Crow South Florida* (Chicago: University of Chicago Press, 2014).

3. Alex Stepick, Guillermo Grenier, Max Castro, and Marvin Dunn, eds., *This Land Is Our Land: Immigrants and Power in Miami* (Berkeley: University of California Press, 2003); Melanie Shell-Weiss, *Coming to Miami: A Social History* (Gainesville: University Press of Florida, 2009).

4. Richard Florida and Steven Pedigo, *Toward a More Inclusive Region: Inequality and Poverty in Greater Miami* (Miami: Miami Urban Future Initiative, 2019), 5; Elyssa Kirkham, "Places Where Adults Still Live with Their Parents," *Magnify Money*, April 2, 2019, https://www.magnifymoney.com/blog/life-events/where-adults-still-live-with-their-parents/. Miami has ranked at the top of the country for millennials living with their parents for several years, according to other real estate websites. Jerry Iannelli, "More Millennials Live with Their Parents in Miami Than Anywhere Else," *Miami New Times*, March 22, 2017, https://www.miaminewtimes.com/news/miami-florida-has-highest-number-of-millennials-living-with-parents-at-home-9220450#:~:text=In%20fact

%2C%20a%20new%20study,national%20average%20of%2034.1%20 percent.

5. RealtyHop, "RealtyHop Housing Affordability Index: February 2022," RealtyHop blog, March 4, 2022, https://www.realtyhop.com/blog/housing-affordability-index-february-2022/.

6. Paul Owers, "Study: Florida Rents Far Outpacing Leasing Trends," Florida Atlantic University News Desk, March 22, 2022, https://www.fau.edu/newsdesk/articles/florida-rents.php.

7. RealtyHop, "RealtyHop Housing Affordability Index."

8. Joshua Ceballos, "'It's Tragic': Camillus House beyond Capacity as Miami Rents Rise," *Miami New Times*, July 1, 2022, https://www.miaminewtimes.com/news/camillus-house-homeless-shelter-exceeds-capacity-as-miami-rents-rise-14788472.

9. Phil Prazan, "Miami-Dade Declares 'Affordability Emergency' as Rents Continue to Skyrocket," *NBC News*, April 8, 2022, https://www.nbcmiami.com/news/local/miami-dade-county-declares-affordability-emergency-as-rents-continue-to-skyrocket/2732370/.

10. Gean Moreno, "Glitch-People," *Affidavit*, November 21, 2016, http://www.affidavit.art/articles/glitch-people.

11. Chiara Bottici, *Imaginal Politics: Images beyond Imagination and the Imaginary* (New York: Columbia University Press, 2014), 1.

12. Bottici, 1.

13. Kevin Grove, Savannah Cox, and Allain Barnett, "Racializing Resilience: Assemblage, Critique, and Contested Futures in Greater Miami Resilience Planning," *Annals of the American Association of Geographers* 110, no. 5 (2020): 1613–30; Kevin Grove, Allain Barnett, and Savannah Cox, "Designing Justice? Race and the Limits of Recognition in Greater Miami Resilience Planning," *Geoforum* 117 (2020): 134–43.

14. Grove, Cox, and Barnett, "Racializing Resilience"; Grove, Barnett, and Cox, "Designing Justice?"; Greater Miami and the Beaches, *Resilient 305*, May 30, 2019, https://www.mbrisingabove.com/wp-content/uploads/Resilient305_final.pdf.

15. Bruce Braun, "Futures: Imagining Socioecological Transformation—An Introduction," *Annals of the American Association of Geographers* 105, no. 2 (2015): 239; Lauren Rickards, Ray Ison, Hartmut Fünfgeld, and John Wiseman, "Opening and Closing the Future: Climate Change, Adaptation, and Scenario Planning," *Environment and Planning C: Government and Policy* 32, no. 4 (2014): 587–602.

16. Naomi Klein, *This Changes Everything: Capitalism vs. the Climate* (New York: Simon & Schuster, 2014); Anna Tsing, *The Mushroom*

at the End of the World: On the Possibility of Life in Capitalist Ruins (Princeton: Princeton University Press, 2016).

17. Julian Reid, "Securing the Imagination: The Politics of the Resilient Self," in *The Resilience Machine*, ed. Jim Bohland, Simin Davoudi, and Jennifer Lawrence (Abingdon: Routledge, 2018), 29–42; Simin Davoudi, "Imagination and Spatial Imaginaries: A Conceptual Framework," in Simin Davoudi, Ruth Raynor, Bryonie Reid, Jenny Crawford, Olivier Sykes, and David Shaw, "Policy and Practice Spatial Imaginaries: Tyrannies or Transformations?," *Town Planning Review* 89, no. 2 (2018): 97–107.

18. Herman Kahn and Anthony J. Weiner, *The Year 2000: A Framework for Speculation on the Next Thirty Years* (New York: Macmillan, 1967); Peter Schwartz, *The Art of the Long View* (New York: Doubleday, 1991); Gill Ringland and Laurie Young, eds., *Scenarios in Marketing: From Vision to Decision* (Chichester: Wiley, 2006).

19. See, e.g., Intergovernmental Panel on Climate Change, *Special Report on Emissions Scenarios* (2000), https://www.ipcc.ch/site/assets/uploads/2018/03/emissions_scenarios-1.pdf; Rickards et al., "Opening and Closing the Future"; Renata Tyszczuk, Joe Smith, and Robert Butler, eds., *Culture and Climate Change: Scenarios* (Cambridge, UK: Shed, 2019); Stephen R. J. Sheppard, Alison Shaw, David Flanders, Sarah Burch, Arnim Wiek, Jeff Carmichael, John Robinson, and Stewart Cohen, "Future Visioning of Local Climate Change: A Framework for Community Engagement and Planning with Scenarios and Visualization," *Futures* 43, no. 4 (2011): 400–412; Noelia Guaita García, Julia Martínez-Fernández, and Carl Fitz, "Environmental Scenario Analysis on Natural and Social-Ecological Systems: A Review of Methods," *Approaches and Applications Sustainability* 12 (2020): 7542; Wood Wade, *Scenario Planning: A Field Guide to the Future* (Hoboken, N.J.: Wiley & Sons, 2012).

20. Zoé A. Hamstead, David M. Iwaniec, Timon McPhearson, Marta Berbés-Blázquez, Elizabeth M. Cook, and Tischa A. Muñoz-Erickson, *Resilient Urban Futures* (Cham: Springer, 2021); Grace Abou Jaoude, Olaf Mumm, and Vanessa Miriam Carlow, "An Overview of Scenario Approaches: A Guide for Urban Design and Planning," *Journal of Planning Literature* 37, no. 3 (2022): 467–87; Robert Goodspeed, *Scenario Planning for Cities and Regions: Managing and Envisioning Uncertain Futures* (Cambridge, Mass.: Lincoln Institute of Land Policy, 2020).

21. Kara Elliott-Ortega, "Urban Design as Problem Solving: Design Thinking in the Rebuild by Design Resiliency Competition" (mas-

ter's thesis, Department of Urban Studies and Planning, Massachusetts Institute of Technology, ,2015), 50.

22. Stephen Collier, Savannah Cox, and Kevin Grove, "Rebuilding by Design in Post-Sandy New York," *Limn* 7 (2016).

23. Luc Boltanski and Eve Chiapello Boltanski, *The New Spirit of Capitalism* (London: Verso, 2005).

24. Kevin Grove, *Resilience* (Abingdon: Routledge, 2018).

25. As Grove, Barnett, and Cox note, residents of Miami's poor and segregated neighborhoods like Overtown have repeatedly been "objects of well-intentioned reformers' attempts to recognize and 'cope' with extreme socio-economic inequality in headline-grabbing, 'innovative' ways." Grove, Barnett, and Cox, "Designing Justice?," 2.

26. Arnab Chakraborty and Andrew McMillan, "Scenario Planning for Urban Planners: Toward a Practitioners Guide," *Journal of the American Planning Association* 81, no. 1 (2015): 18–29.

27. Timon McPhearson, Christopher M. Raymond, Natalie Gulsrud, Christian Albert, Neil Coles, Nora Fagerholm, Michiru Nagatsu, Anton Stahl Olafsson, Niko Soininen, and Kati Vierikko, "Radical Changes Are Needed for Transformations to a Good Anthropocene," *npj Urban Sustainability* 1, no. 5 (2021); David M. Iwaniec, Elizabeth M. Cook, Melissa J. Davidson, Marta Berbés-Blázquez, Matei Georgescu, E. Scott Krayenhoff, Ariane Middel, David A. Sampson, and Nancy B. Grimm, "The Co-production of Sustainable Future Scenarios," *Landscape and Urban Planning* 197 (2020): 103744; Per Olsson, Victor Galaz, and Wiebren J. Boonstra, "Sustainability Transformations: A Resilience Perspective," *Ecology and Society* 19, no. 4 (2015).

28. Miami has been the focus of diverse future scenario visioning and planning activities, including, among others, Miami-Dade Urban Long Term Research Area (ULTRA), https://miamiultra.fiu.edu/about-us/; Greater Miami and the Beaches, *Resilient 305;* City of Miami Beach, "Inspirational Visioning Workshop for a Resilient Miami Beach in 2070," July 21, 2021, https://www.mbrisingabove.com/your-city-at-work/resilience-strategy/visioning-workshops/; and Harvard Graduate School of Design, Future of the American City Design Studio, https://fotac.gsd.harvard.edu/learn.

29. Urban Resilience to Extreme Events Sustainability Research Network (URExSRN), *Envisioning Future Urban Resilience to Climate Change in the Greater Miami Area*, workshop report, April 12, 2019, https://sustainability-innovation.asu.edu/urbanresilience/wp-content/

uploads/sites/4/2020/02/greater-miami-area-scenarios-workshopreport.pdf.

30. Lauren Rickards, John Wiseman, Taegen Edwards, and Che Biggs, "The Problem of Fit: Scenario Planning and Climate Change Adaptation in the Public Sector," *Environment and Planning C* 32, no. 4 (2014): 641–62; Mark Neocleous, "Resisting Resilience," *Radical Philosophy* 178 (March/April 2013): 2–7; Claudia Aradau, "The Myth of Preparedness," *Radical Philosophy* 161 (May/June 2010): 1–7.

31. Iwaniec et al., "Co-production of Sustainable Future Scenarios," 3. This scenario methodology was used in the URExSRN's *Envisioning Future Urban Resilience to Climate Change in the Greater Miami Area* workshop, which yielded the above-mentioned future visions.

32. Aradau, "The Myth of Preparedness."

33. James C. Scott, *Seeing Like a State: How Certain Schemes to Improve the Human Condition Have Failed* (New Haven: Yale University Press, 1998).

34. M. Christine Boyer, *Dreaming the Rational City: The Myth of American City Planning* (Cambridge: MIT Press, 1983).

35. Fikret Berkes and Carl Folke, eds., *Linking Social and Ecological Systems: Management Practices and Social Mechanisms for Building Resilience* (Cambridge, UK: Cambridge University Press, 1998); Elinor Ostrom, "A General Framework for Analyzing Sustainability of Social-Ecological Systems," *Science* 325, no. 5939 (2009): 419–42; C. S. Holling, "Resilience and Stability of Ecological Systems," *Annual Review of Ecology and Systematics* 4 (1973): 1–23.

36. Scott, *Seeing Like a State*; Warren Magnusson, *Politics of Urbanism: Seeing Like a City* (Abingdon: Routledge, 2012); Ash Amin and Nigel Thrift, *Seeing Like a City* (Cambridge, UK: Polity, 2016).

37. Amin and Thrift, *Seeing Like a City*, 29.

38. Amin and Thrift, 25.

39. Stephanie Wakefield, "Urban Resilience as Critique: Problematizing Infrastructure in Post-Sandy New York City," *Political Geography* 79 (2020): 102148.

40. Oliver Wainwright, "Bjarke Ingels on the New York Dryline: 'We think of it as the love-child of Robert Moses and Jane Jacobs,'" *Guardian*, March 9, 2015, https://www.theguardian.com/cities/2015/mar/09/bjarke-ingels-new-york-dryline-park-flood hurricane-sandy.

41. Aradau, "The Myth of Preparedness," 7.

42. Jeroen van der Heijden, Chiara Certomà, and Harriet Bulkeley, "The Politics of Urban Climate Futures: Recognition, Experimentation,

Orchestration," in *Urban Climate Politics: Agency and Empowerment*, ed. Jeroen van der Heijden, Harriet Bulkeley, and Chiara Certomà (Cambridge: Cambridge University Press, 2019), 231–42; Grove, *Resilience*.

43. Reid, "Securing the Imagination."

44. Benedict Anderson, *Imagined Communities* (London: Verso Books, 2016).

45. Michel de Certeau, *The Practice of Everyday Life* (Berkeley: University of California Press, 2011).

46. Derek Gregory, "Geographical Imagination," in *The Dictionary of Human Geography*, ed. Derek Gregory, Ron Johnston, Geraldine Pratt, Michael Watts, Sarah Whatmore, 5th ed. (Malden, Mass.: Wiley-Blackwell, 2009), 284.

47. Frédéric Neyrat, *The Unconstructable Earth: An Ecology of Separation* (New York: Fordham University Press, 2019); Bruno Latour, *Down to Earth: Politics in a New Climatic Regime* (Cambridge, UK: Polity Press, 2018); Donna Haraway, *Staying with the Trouble: Making Kin in the Chthulucene* (Durham: Duke University Press, 2016); Tsing, *Mushroom*.

48. Jen Jack Gieseking, "Geographical Imagination," in *International Encyclopedia of Geography*, ed. Douglas Richardson, Noel Castree, Michael F. Goodchild, Audrey Kobayashi, Weidong Liu, and Richard A. Marston (New York: Wiley-Blackwell and the Association of American Geographers, 2017). For critical feminist geographical work on the view from above, see, e.g., Mei-Po Kwan, "Feminist Visualization: Re-envisioning GIS as a Method in Feminist Geographic Research," *Annals of the Association of American Geographers* 92, no. 4 (2002): 645–61; Donna Haraway, *Simians, Cyborgs, and Women: The Reinvention of Nature* (New York: Routledge, 1991); Derek Gregory, *Geographical Imaginations* (Malden, Mass.: Wiley-Blackwell, 1994); Rosalyn Deutsche, "Boys Town," *Environment and Planning D: Society and Space* 9, no. 1 (1991): 5–30.

49. Thaddeus Pawlowski, "CRCL Attends South Florida Serious Games," Columbia University Center for Resilient Cities and Landscapes, July 1, 2018, https://crcl.columbia.edu/news/crcl-attends-south-florida-serious-games; Florida Disaster Resilience Initiative, https://southflorida.resiliencesystem.org/.

50. Raymond A. Mohl, "Whitening Miami: Race, Housing, and Government Policy in Twentieth-Century Dade County," *Florida Historical Quarterly* 79, no. 3 (2001): 320; Paul S. George, "Colored Town:

Miami's Black Community, 1896–1930," *Florida Historical Quarterly* 56, no. 4 (1978): 432–47.

51. Marvin Dunn, *Black Miami in the Twentieth Century* (Gainesville: University Press of Florida, 2016).

52. Raymond A. Mohl, "The Origins of Miami's Liberty City," *Florida Environmental and Urban Issues* (July 1985): 11.

53. Mohl, "Whitening Miami."

54. Mohl.

55. Connolly, *A World More Concrete*, 14.

56. Grove, Barnett, and Cox, "Designing Justice?"

57. Edward Murray and Kevin T. Greiner, *Liberty City: Economic Analysis and Opportunities Report*, South Florida Housing Studies Consortium (Miami-Dade County, Florida International University Metropolitan Center, University of Miami Office of Civic and Community Engagement) (2017), 2.

58. Southeast Overtown/Park West Community Redevelopment Agency, *Assessment of Need Study*, 2018, 15, http://www.miamicra.com/seopwcra/docs/SEOPW%20CRA%20FON%202018.pdf.

59. Alejandro Portes and Ariel C. Armony, *The Global Edge: Miami in the 21st Century* (Oakland: University of California Press, 2018), xvi.

60. Andy Merrifield, *The Politics of the Encounter: Urban Theory and Protest under Planetary Urbanization* (Athens: University of Georgia Press, 2013), 8; Sallie A. Marston, "The Social Construction of Scale," *Progress in Human Geography* 24, no. 2 (June 2000): 219–42.

61. Latour, *Down to Earth*.

62. Zac Taylor and Manuel Aalbers, "Climate Gentrification: Risk, Rent, and Restructuring in Greater Miami," *Annals of the American Association of Geographers* 12, no. 6 (2022): 1685–1701; See also Jesse Keenan, Thomas Hill, and Anurag Gumber, "Climate Gentrification: From Theory to Empiricism in Miami-Dade County, Florida," *Environmental Research Letters* 13, no. 5 (2018): 054001.

63. Miami Climate Alliance, *Housing Justice Is Climate Justice*, 2020, https://d3n8a8pro7vhmx.cloudfront.net/catalystmiami/pages/140/attachments/original/1590720073/Housing_Justice_is_Climate_Justice_2020-compressed.pdf; Savannah Cox, Kevin Grove, and Allain Barnett, "Design-Driven Resilience and the Limits of Geographic Critique," *The Geographical Journal*, 188, no. 2 (2022): 294–308.

64. Brad Evans and Julian Reid, *Resilient Life: The Art of Living Dangerously* (London: Polity, 2014); Ben Anderson, Kevin Grove, Lauren Rickards, and Matthew Kearnes, "Slow Emergencies: Temporality

and the Racialized Biopolitics of Emergency Governance," *Progress in Human Geography* 44, no. 4 (2020): 621–39; Lauren Berlant, *Cruel Optimism* (Durham: Duke University Press, 2011).

65. David Chandler, *Ontopolitics in the Anthropocene: An Introduction to Mapping, Sensing, and Hacking* (London: Routledge, 2018).

66. Eric Klinenberg, *Palaces for the People: How Social Infrastructure Can Help Fight Inequality, Polarization, and the Decline of Civic Life* (New York: Crown Publishing Group, 2019). Alan Latham and Jack Layton, "Social Infrastructure and the Public Life of Cities: Studying Urban Sociality and Public Spaces," *Geography Compass* 13, no. 7 (2019): e12444.

67. AbdouMalique Simone, "People as Infrastructure: Intersecting Fragments in Johannesburg," *Public Culture* 16, no. 3 (2004): 407–29; AbdouMalique Simone, "Ritornello: 'People as Infrastructure,'" *Urban Geography* 42, no. 9 (2021): 1341–48. See also Colin McFarlane and Jonathan Silver, "Navigating the City: Dialectics of Everyday Urbanism," *Transactions of the Institute of British Geographers* 42 (2017): 458–71; Jean-Paul Addie, "Urban Life in the Shadows of Infrastructural Death: From People as Infrastructure to Dead Labor and Back Again," *Urban Geography* 42, no. 9 (2021): 1349–61.

68. Stephen Collier and Andrew Lakoff, "The Vulnerability of Vital Systems: How 'Critical Infrastructure' Became a Security Problem," in *Securing the Homeland: Critical infrastructure, Risk and (In)security*, ed. Miriam Anna Dunn and Kristian Søby Kristensen (London: Routledge, 2008), 17–39; Department of Homeland Security, *Lessons Learned: Social Media and Hurricane Sandy*, Virtual Social Media Working Group and DHS First Responders Group, 2013, https://www.dhs.gov/sites/default/files/publications/Lessons%20Learned%20Social%20Media%20and%20Hurricane%20Sandy.pdf.

69. Federal Emergency Management Agency, *A Whole Community Approach to Emergency Management: Principles, Themes, and Pathways for Action*, FEMA FDOC 104-008-1, 2011, https://www.fema.gov/sites/default/files/2020-07/whole_community_dec2011__2.pdf.

70. Eric Klinenberg, "Toward a Stronger Social Infrastructure: A Conversation with Eric Klinenberg," *Urban Omnibus*, 2013, https://urbanomnibus.net/2013/10/toward-a-stronger-social-infrastructure-a-conversation-with-eric-klinenberg/; Andrea McArdle, "Storm Surges, Disaster Planning, and Vulnerable Populations at the Urban Periphery: Imagining a Resilient New York after Superstorm Sandy," *Idaho Law Review* 50 (2014): 19–47.

71. Urban Land Institute, *After Sandy: Advancing Strategies for Long-Term Resilience and Adaptability* (Washington, D.C.: Urban Land Institute, 2013): 33; Eric Klinenberg, "Adaptation," *New Yorker,* December 30, 2012, https://www.newyorker.com/magazine/2013/01/07/adaptation-eric-klinenberg; "first social media disaster": Joe Coscarelli, "Hurricane Sandy: A Perfect Social Media Storm," *New York Magazine*, October 30, 2012, http://nymag.com/intelligencer/2012/10/hurricane-sandy-perfect-social-media-storm.html; Department of Homeland Security, *Lessons Learned*, 13–14.

72. David Chandler and Julian Reid, *The Neoliberal Subject: Resilience, Adaptation, and Vulnerability* (London: Rowman & Littlefield International, 2016).

73. Federal Emergency Management Agency, *A Whole Community Approach.*

74. Shari Holbert Lipner, "CERTified Help," *Miami Beach Magazine*, Winter 2019, 49.

75. Nikolas Rose, "The Death of the Social? Re-figuring the Territory of Government," *Economy and Society* 25, no. 3 (1996): 327–56.

76. Simone, "Ritornello," 1342.

77. Simone, 1343.

78. Ash Amin, "Surviving the Turbulent Future," *Environment and Planning D*, 31, no. 1 (2013): 140–56.

79. Addie, "Urban Life in the Shadows."

80. Reid, "Securing the Imagination," 29–30.

81. Bjørn Enge Bertelsen, "A Lesser Human? Utopian Registers of Urban Reconfiguration in Maputo, Mozambique," *Social Anthropology/Anthropologie Sociale* 29, no. 1 (2021): 90.

82. Evans and Reid, *Resilient Life.*

83. Tsing, *Mushroom.*

84. Erik Swyngedouw, "Depoliticized Environments and the Promises of the Anthropocene," in *The International Handbook of Political Ecology*, ed. Raymond L. Bryan (Cheltenham: Edward Elgar Publishing Limited, 2015), 136.

85. Reid, "Securing the Imagination," 36.

86. Reid.

87. Iwaniec et al., "Co-production of Sustainable Future Scenarios," 1.

88. See https://littlerivercooperative.com.

89. Melissa Bernardo, "Miami's Instagram Farmers: Increasing

Actually-Existing Resilience in Response to Multidimensional Crises," Dimensions of Political Ecology Annual Conference, February 20, 2021.

90. EcoTech Visions Foundation, Annual Report, September, 12, 2018, Miami, 39.

91. Miami Climate Alliance, *Housing Justice Is Climate Justice*.

92. Stephanie Wakefield, *Anthropocene Back Loop: Experimentation in Unsafe Operating Space* (London: Open Humanities Press, 2020).

93. Grove, *Resilience*, 266; see also Kevin Grove, "Adaptation Machines and the Parasitic Politics of Life in Jamaican Disaster Resilience," *Antipode* 46, no. 3 (2014): 611–28.

94. Grove, *Resilience*, 263.

3. Flows against Flows

1. Christopher Flavelle, "Miami Will Be Underwater Soon. Its Drinking Water Could Go First," *Bloomberg*, August 29, 2018, https://www.bloomberg.com/news/features/2018-08-29/miami-s-other-water-problem; Jayantha Obeysekera, Wendy Graham, Michael Sukop, Tirusew Asefa, Dingbao Wang, Kebreab Ghebremichael, and B. M. Mwashote, "Implications of Climate Change on Florida's Water Resources," in *Florida's Climate: Changes, Variations, & Impacts*, ed. Eric P. Chassignet, James W. Jones, Vasubandhu Misra, and Jayantha Obeysekera (Gainesville: Florida Climate Institute, 2017), 83–124; Scott T. Prinos, "Map of the Approximate Inland Extent of Saltwater at the Base of the Biscayne Aquifer in Miami-Dade County, Florida, 2018," U.S. Geological Survey, Scientific Investigations Map 3438, pamphlet, 2019, https://doi.org/10.3133/sim3438; Frederick Bloetscher, Barry Heimlich, and Daniel E. Meeroff, "Development of an Adaptation Toolbox to Protect Southeast Florida Water Supplies from Climate Change," *Environmental Reviews* 19 (December 2011): 397–417.

2. Randall Parkinson, Peter Harlem, and John Meeder, "Managing the Anthropocene Marine Transgression to the Year 2100 and Beyond in the State of Florida U.S.A.," *Climactic Change* 128, nos. 1–2 (2015): 85–98; John Meeder and Randall Parkinson, "SE Saline Everglades Transgressive Sedimentation in Response to Historic Acceleration in Sea-Level Rise: A Viable Marker for the Base of the Anthropocene?," *Journal of Coastal Research* 34, no. 2 (2018): 490–97.

3. Daniel L. Childers, Evelyn E. Gaiser, and Laura A. Ogden, eds., *The Coastal Everglades: The Dynamics of Social-Ecological Transfor-*

mation in the South Florida Landscape (New York: Oxford University Press, 2019), 9.

4. Dan Weissmann, "Rising Seas Threaten South Florida's Drinking Water," *Marketplace*, February 10, 2015, https://www.marketplace.org/2015/02/10/rising-seas-threaten-south-floridas-drinking-water/.

5. Gary R. Mormino and Raymond Arsenault, foreword to *Gladesmen: Gator Hunters, Moonshiners, and Skiffers*, by Glenn Simmons and Laura Ogden (Tallahassee: University of Florida Press, 1998), x.

6. Marjory Stoneman Douglas, *The Everglades: River of Grass* (Sarasota, Fla.: Pineapple Press, 2007).

7. Mike Vogel, "Restoring the Florida Everglades: Where Things Stand," *Florida Trend*, January 21, 2021, https://www.floridatrend.com/article/30605/restoring-the-florida-everglades-where-things-stand.

8. U.S. Army Corps of Engineers, "South Florida Ecosystem Restoration (SFER) Program Overview," 2018, https://usace.contentdm.oclc.org/utils/getfile/collection/p16021coll11/id/2552; "Restoring America's Everglades," U.S. Department of the Interior, https://www.evergladesrestoration.gov/.

9. Paul Julian and Khandker Ishtiaq, "Getting the Water Right," *The Everglades Foundation Science Insider* 87 (Summer 2023): 27.

10. Prinos, "Map of the Approximate Inland Extent," 2.

11. Jessica Cattelino, "The Cultural Politics of Water in the Everglades and Beyond," *Hau: Journal of Ethnographic Theory* 5, no. 3 (2015): 235–50; Laura Ogden, *Swamplife: People, Gators, and Mangroves Entangled in the Everglades* (Minneapolis: University of Minnesota Press, 2011).

12. Laura Ogden, "The Everglades Ecosystem and the Politics of Nature," *American Anthropologist* 110, no. 1 (2008): 22. See also Michael Grunwald, *The Swamp: The Everglades, Florida, and the Politics of Paradise* (New York: Simon & Schuster, 2006).

13. David McCally, *Everglades: An Environmental History* (Gainesville: University Press of Florida, 1999).

14. C. S. Holling, "The Structure and Dynamics of the Everglades System: Guidelines for Ecosystem Restoration," in *Everglades: The Ecosystem and Its Restoration*, ed. Steven Davis and John Ogden (Delray Beach, Fla.: St. Lucie Press, 1994), 741–56; Lance Gunderson and Stephen S. Light, "Adaptive Management and Adaptive Governance in the Everglades Ecosystem," *Policy Sciences* 39, no. 4 (2006): 323–34.

15. Grunwald, *The Swamp*, 3.

16. Christopher Joyce, "Rising Seas Push Too Much Salt into the Florida Everglades," *NPR,* May 25, 2016, https://www.npr.org/2016/05/25/477014085/rising-seas-push-too-much-salt-into-the-florida-everglades.

17. Hannah Northey, "Restoration Costs Soar as Seas Rise and Pythons Slither In," *E&E News,* January 4, 2021, https://www.eenews.net/stories/1063721687.

18. See, e.g., The Everglades Foundation, "The Everglades and Climate Change," https://www.evergladesfoundation.org/climate-change.

19. Joyce, "Rising Seas."

20. Flavelle, "Miami Will Be Underwater Soon."

21. Cattelino "Cultural Politics," 235–36.

22. Matthew Gandy, *Concrete and Clay: Reworking Nature in New York City* (Cambridge: MIT Press, 2002).

23. Neil Brenner, ed., *Implosions/Explosions: Towards a Study of Planetary Urbanization* (Berlin: Jovis, 2014), 20.

24. Ashley Carse, "Keyword: Infrastructure: How a Humble French Engineering Term Shaped the Modern World," in *Infrastructures and Social Complexity: A Companion*, ed. Penelope Harvey, Casper Jensen, and Atsuro Mori (Abingdon: Routledge, 2017), 27.

25. See Stephen Collier and Andrew Lakoff, "The Vulnerability of Vital Systems: How 'Critical Infrastructure' Became a Security Problem," in *Securing the Homeland: Critical Infrastructure, Risk, and (In)security*, ed. Miriam Anna Dunn and Kristian Søby Kristensen (London: Routledge, 2008), 17–39; Tony H. Grubesic and Alan T. Murray, "Vital Nodes, Interconnected Infrastructures, and the Geographies of Network Survivability," *Annals of the Association of American Geographers* 96, no. 1 (2006): 64–83; Jochen Monstadt and Martin Schmidt, "Urban Resilience in the Making? The Governance of Critical Infrastructures in German Cities," *Urban Studies* 56, no. 11 (2019): 2353–71; Jon Coaffee and Jonathan Clarke, "Critical Infrastructure Lifelines and the Politics of Anthropocentric Resilience," *Resilience* 5, no. 3 (2016): 1–21; Steve Graham and Simon Marvin, *Splintering Urbanism: Networked Infrastructures, Technological Mobilities, and the Urban Condition* (London: Routledge, 2001).

26. See, e.g., Bruce Braun, "Environmental Issues: Writing a More-than-Human Urban Geography," *Progress in Human Geography* 29, no. 5 (2005): 635–50; Deborah Cowen, *The Deadly Life of Logistics: Mapping Violence in Global Trade* (Minneapolis: University of Minnesota

Press, 2014); Alex Hughes and Suzanne Reimer, eds., *Geographies of Commodity Chains* (London: Routledge, 2004).

27. Bruce Braun, "A New Urban Dispositif? Governing Life in an Age of Climate Change," *Environment and Planning D* 32, no. 1 (2014): 49–64.

28. Swati Chattopadhyay, *Unlearning the City: Infrastructure in a New Optical Field* (Minneapolis: University of Minnesota Press, 2012), x. On the "promise" of modern infrastructure, see Nikhil Anand, Akhil Gupta, and Hannah Appel, eds., *The Promise of Infrastructure* (Durham: Duke University Press, 2018); Kregg Hetherington, "Surveying the Future Perfect: Anthropology, Development and the Promise of Infrastructure," in Harvey, Jensen, and Mori, *Infrastructures and Social Complexity: A Companion*, 40–51. See also Bruce Braun and Noel Castree, *Social Nature: Theory, Practice and Politics* (London: Wiley-Blackwell, 2001); Graham and Marvin, *Splintering Urbanism*; Neil Smith, *Uneven Development: Nature, Capital, and the Production of Space* (Athens: University of Georgia Press, 1990); Stephanie Wakefield, "Infrastructures of Liberal Life: From Modernity and Progress to Resilience and Ruins," *Geography Compass* 12, no. 7 (2018): e12377.

29. Stephanie Wakefield, *Anthropocene Back Loop: Experimentation in Unsafe Operating Space* (London: Open Humanities Press, 2020); Simon Dalby, "Biopolitics and Climate Security in the Anthropocene," *Geoforum* 49 (2013): 184–92; Christoph Rosol, Sara Nelson, and Jürgen Renn, "Introduction: In the Machine Room of the Anthropocene," *Anthropocene Review* 4, no. 1 (2017): 2–8.

30. Kara Elliott-Ortega, "Urban Design as Problem Solving: Design Thinking in the Rebuild by Design Resiliency Competition" (master's thesis, Department of Urban Studies and Planning, Massachusetts Institute of Technology, 2015), 83.

31. Stephanie Wakefield, "Urban Resilience as Critique: Problematizing Infrastructure in Post-Sandy New York City," *Political Geography* 79 (2020): 102148.

32. Douglas, *The Everglades*, 286.

33. Holling, "Structure and Dynamics;" Gunderson and Light, "Adaptive Management."

34. Urban Land Institute, *Stormwater Management and Climate Adaptation Review: Urban Land institute Advisory Services Panel Report* (Washington, D.C.: Urban Land Institute, 2018), 18.

35. Hillary Angelo, *How Green Became Good: Urbanized Nature*

and the Making of Cities and Citizens (Chicago: University of Chicago Press, 2021); Dorceta E. Taylor, "Central Park as a Model for Social Control: Urban Parks, Social Class and Leisure Behavior in Nineteenth-Century America," *Journal of Leisure Research* 31, no. 4 (1999): 420–77; Scott M. Larson, "Imagining Social Justice and the False Promise of Urban Park Design," *Environment and Planning A: Economy and Space* 50, no. 2 (2018): 391–406.

36. José Maria Cardoso da Silva and Emily Wheeler, "Ecosystems as Infrastructure," *Perspectives in Ecology and Conservation* 15, no. 1 (2017): 32–35; United Nations Educational Scientific and Cultural Organisation, International Co-ordinating Council of the Programme on Man and the Biosphere, Report no. 57: "International Experts Meeting on Ecological Approaches to Urban Planning" (September 24–30, 1984).

37. Richard T. T. Forman and Michel Godron, *Landscape Ecology* (New York: Wiley, 1986).

38. Karl-Friedrich Schrieiber, ed., "Connectivity in Landscape Ecology," Proceedings of the 2nd International Seminar of the International Association for Landscape Ecology (Paderborn: Ferdinand Schöningh, 1988); AJV Selm AJV, "Ecological Infrastructure: A Conceptual Framework for Designing Habitat Networks," in *Connectivity in Landscape Ecology*, ed. Karl-Friedrich Schrieiber, Proceedings of the 2nd International Seminar of the International Association for Landscape Ecology (Paderborn: Ferdinand Schöningh, 1988), 63–66; A. J. van Selm, "Ecological Infrastructure on a National Scale?," in Schrieiber, *Connectivity in Landscape Ecology*, 177–78.

39. Kongjian Yu, Sisi Wang, and Dihua Li, "The Negative Approach to Urban Growth Planning of Beijing, China," *Journal of Environmental Planning and Management* 54, no. 9 (2011): 1209–36.

40. Robert Costanza, Ralph d'Arge, Rudolf de Groot, Stephen Farber, Monica Grasso, Bruce Hannon, Karin Limburg, Shahid Naeem, Robert V. O'Neill, Jose Paruelo, Robert G. Raskin, Paul Sutton, and Marjan van den Belt, "The Value of the World's Ecosystem Services and Natural Capital," *Nature* 387, no. 6630 (1997): 253–60; Gretchen Cara Daily, *Nature's Services: Societal Dependence on Natural Ecosystems* (Washington, D.C.: Island Press, 1997); Millennium Ecosystem Assessment, *Ecosystems and Human Well-Being: Synthesis* (Washington, D.C.: Island Press, 2005); Morgan Robertson, "The Nature That Capital Can See: Science, State, and Market in the Commodification of Ecosystem Services," *Environment and Planning D: Society and Space*

24, no. 3 (2006): 367–87; Morgan Robertson, "The Neoliberalization of Ecosystem Services: Wetland Mitigation Banking and Problems in Environmental Governance," *Geoforum* 35, no. 3 (2004): 361–73; Sara H. Nelson, "Beyond the Limits to Growth: Ecology and the Neoliberal Counterrevolution," *Antipode* 47, no. 2 (2015): 461–80.

41. Karen Bakker, "The Limits of 'Neoliberal Natures': Debating Green Neoliberalism," *Progress in Human Geography* 34, no. 6 (2010): 715–35; Nik Heynen and Paul F. Robbins, "The Neoliberalization of Nature: Governance, Privatization, Enclosure, and Valuation," *Capitalism, Nature, Socialism* 16, no. 1 (2005): 5–8; Cindi Katz, "Whose Nature, Whose Culture? Private Productions of Space and the Preservation of Nature," in *Remaking Reality: Nature at the End of the Millennium*, ed. Bruce Braun and Noel Castree (New York: Routledge, 1998), 46–63; Noel Castree, "Neoliberalising Nature: The Logics of Deregulation and Reregulation," *Environment and Planning A* 40, no. 1 (2008): 131–52; Noel Castree, "Neoliberalising Nature: Processes, Effects, and Evaluations," *Environment and Planning A* 40, no. 1 (2008): 153–73.

42. Meenakshi Jerath, Mahadev Bhat, Victor H. Rivera-Monroy, Edward Castañeda-Moya, Marc Simard, and Robert R. Twilley, "The Role of Economic, Policy, and Ecological Factors in Estimating the Value of Carbon Stocks in Everglades Mangrove Forests, South Florida, USA," *Environmental Science & Policy*, 66 (2016): 160–69.

43. Sara H. Nelson and Patrick Bigger, "Infrastructural Nature," *Progress in Human Geography* 46, no. 1 (2022): 86–107.

44. Robert I. McDonald, *Conservation for Cities: How to Plan and Build Natural Infrastructure* (Washington, D.C.: Island Press, 2015).

45. Pierre Bélanger, "Landscape as Infrastructure," *Landscape Journal* 28, no. 1 (2009): 79–95; Pierre Bélanger, "Infrastructural Ecologies: Fluid, Biotic, Contingent," in *Landscape Infrastructure: Case Studies by SWA*, ed. Charles Waldheim, Gerdo Aquino, and Ying-Yu Hung (Basel: Birkhauser, 2013), 20–27.

46. McDonald, *Conservation for Cities*.

47. Christopher Alexander, "A City Is Not a Tree," *Architectural Forum* 122, no. 1 (April 1965): 58–62 (Part I); 122, no. 2 (May 1965): 58–62 (Part II).

48. Ian McHarg, *Design with Nature* (Garden City, N.Y.: Natural History Press, 1969), 5.

49. Jane Jacobs, *The Death and Life of Great American Cities* (New York: Random House and Vintage Books, 1961).

50. Nelson and Bigger, "Infrastructural Nature."

51. Stephanie Wakefield, "Making Nature into Infrastructure: The Construction of Oysters as a Risk Management Solution in New York City," *Environment and Planning E: Nature and Space* 3, no. 3 (2020): 761–85; Stephanie Wakefield and Bruce Braun, "Oystertecture: Infrastructure, Profanation, and the Sacred Figure of the Human," in *Infrastructure, Environment, and Life in the Anthropocene*, ed. Kregg Hetherington (Durham: Duke University Press, 2019), 193–215.

52. Costanza et al., "Value of the World's Ecosystem."

53. Nelson and Bigger, "Infrastructural Nature."

54. Michel Foucault, *The Birth of Biopolitics: Lectures at the Collège de France, 1978–1979* (New York: Palgrave Macmillan, 2008).

55. Braun, "A New Urban Dispositif?," 58; see also Michael Dillon and Julian Reid, *The Liberal Way of War: Killing to Make Life Live* (London: Routledge, 2009); Stephanie Wakefield, "Infrastructures of Liberal Life: From Modernity and Progress to Resilience and Ruins," *Geography Compass* 12, no. 7 (2018): e12377.

56. Ben Anderson, "Preemption, Precaution, Preparedness: Anticipatory Action and Future Geographies," *Progress in Human Geography* 34, no. 6 (2010): 777–98.

57. Henrik Ernstson and Erik Swyngedouw, "Politicising the Urban Environment," in *Urban Political Ecology in the Anthropo-obscene*, ed. Henrik Ernstson and Erik Swyngedouw (Abingdon: Routledge, 2019), 1–21.

58. Smith, *Uneven Development*.

59. See https://storymaps.arcgis.com/stories/dfd3e4261602415683 015a919dfbafec; Stephen S. Light and J. W. Dineen, "Water Control in the Everglades: A Historical Perspective," in Davis and Ogden, *Everglades*, 47–84; John C. Ogden, Steven M. Davis, and Laura A. Brandt, "Science Strategy for a Regional Ecosystem Monitoring and Assessment Program: The Florida Everglades Example," in *Monitoring Ecosystems: Interdisciplinary Approaches For Evaluating Ecoregional Initiatives*, ed. David E. Busch and Joel C. Trexler (Washington, D.C.: Island Press, 2003), 135–67.

60. Obeysekera et al., "Implications of Climate Change," 4.

61. Cattelino, "Cultural Politics."

62. Michelle Navakas, *Liquid Landscape: Geography and Settlement at the Edge of Early America* (Philadelphia: University of Pennsylvania Press, 2018).

63. John F. Meeder, Randall W. Parkinson, Pablo L. Ruiz, and Michael S. Ross, "Saltwater Encroachment and Prediction of Future Eco-

system Response to the Anthropocene Marine Transgression, Southeast Saline Everglades, Florida," *Hydrobiologia* 803 (2017): 29–48; Light and Dineen, "Water Control in the Everglades." Childers, Gaiser, and Ogden, *The Coastal Everglades*, 18.

64. Joyce, "Rising Seas."

65. Jan Sjostrom, "Experts Advocate for Everglades Restoration at the Four Arts," *Palm Beach Daily News*, December 19, 2016.

66. Martha Musgrove, "Will Scott's Tantrum Threaten Everglades Restoration?," *Sun Sentinel*, September 24, 2015.

67. Allen G. Breed, "Scientists Work to Reverse 'War on the Ecosystem' in Everglades," *Christian Science Monitor*, December 3, 2019, https://www.csmonitor.com/Environment/2019/1203/Scientists-work-to-reverse-war-on-the-ecosystem-in-Everglades.

68. Northey, "Restoration Costs Soar."

69. The South Florida Ecosystem Restoration Task Force, "Comprehensive Everglades Restoration Plan (CERP)," http://evergladesrestoration.gov.

70. Breed, "Scientists Work."

71. Wakefield, "Making Nature into Infrastructure."

72. Wakefield, "Making Nature into Infrastructure"; Wakefield and Braun, "Oystertecture."

73. Braun, "New Urban Dispositif?," 59.

74. Jane Bennett, *Vibrant Matter: A Political Ecology of Things* (Durham: Duke University Press, 2007); Bruno Latour, *We Have Never Been Modern* (Cambridge, Mass.: Harvard University Press, 1993).

75. Anna Tsing, *The Mushroom at the End of the World: On the Possibility of Life in Capitalist Ruins* (Princeton: Princeton University Press, 2016), 1.

76. Timothy Luke, "On Environmentality: Geo-Power and Eco-Knowledge in the Discourses of Contemporary Environmentalism," *Cultural Critique*, 31, no. 2 (1995): 57–81.

77. Ogden, "The Everglades Ecosystem." For Ogden, "envisioning the Everglades as a water management system is one way of collapsing the messiness of regional environmental and social relationships into a set of technological problems that have imagined solutions. . . . In this case, the Everglades is imagined as an ecological landscape overlaid with a water management system, disarticulated from the social and natural processes that continue to reshape it" ("The Everglades Ecosystem," 28). Everglades as infrastructure pushes this further, imagining the *integration* of ecological landscape, water management system, and

constitutive social and natural processes. Imagined here is that Everglades water flows, once restored, would not only be manageable *via* technical solutions but that they would *themselves* act technically, as an infrastructure. Here CERP is no longer imagined simply as a better means of managing the existing regional water system, but also as a means of managing other water systems—the intrusion of salt water—and their effects on urban life.

78. Cattelino, "Cultural Politics," 246.

79. Quoted in Kimberly Miller, "ANALYSIS: Why Everglades Restoration Is Losing to Sea Level Rise," *Palm Beach Post*, February 15, 2018, https://www.palmbeachpost.com/story/news/2018/02/16/analysis-why-everglades-restoration-is/6890046007/.

80. David Chandler and Julian Reid, *Becoming Indigenous: Governing Imaginaries in the Anthropocene* (London: Rowman & Littlefield International, 2019).

81. Benjamin Bratton, *The Stack: On Software and Sovereignty* (Cambridge: MIT Press, 2016).

82. Gandy, *Concrete and Clay*; Maria Kaika, *City of Flows: Modernity, Nature, and the City* (New York: Routledge, 2005).

83. Ross Exo Adams, "Natura Urbans, Natura Urbanata: Ecological Urbanism, Circulation, and the Immunization of Nature," *Environment and Planning D: Society and Space* 32, no. 1 (2014): 12–29.

84. Ross Exo Adams makes a similar argument regarding urban resilience and urbanization broadly speaking. See Adams, "Natura Urbans," 26.

85. Braun, "A New Urban Dispositif?," 58.

86. Obeysekera et al., "Implications of Climate Change," 84.

87. Latour, *We Have Never Been Modern*.

88. National Academies of Sciences, Engineering, and Medicine, *Progress toward Restoring the Everglades: The Seventh Biennial Review* (Washington, D.C.: National Academies Press, 2018), 10.

89. Angela Nicoletti, "The Building Block of the Everglades Is in Danger," *FIU News*, February 25, 2020, https://news.fiu.edu/2020/the-building-block-of-the-everglades-is-in-danger.

90. National Academies of Sciences, Engineering, and Medicine, *Progress toward Restoring the Everglades*, xiv.

91. Shimelis Dessu, René Price, Tiffany Troxler, and John Kominoski, "Effects of Sea-Level Rise and Freshwater Management on Long-Term Water Levels and Water Quality in the Florida Coastal

Everglades," *Journal of Environmental Management* 211, no. 1 (2018): 164–76.

92. Nicoletti, "The Building Block."

93. Geologist Randall Parkinson quoted in Oliver Milman, "Everglades under Threat as Florida's Mangroves Face Death by Rising Sea Level," *Guardian*, May 2, 2018, https://www.theguardian.com/us-news/2018/may/02/mangroves-everglades-florida-rising-sea-level.

94. Mike Ross, S. Stoffella, J. Meeder, L. Scinto, R. Vidales, H. Biswas, S. Charles, A. Wachnicka, and J. Sah, "Ecosystem Dynamics in the White Zone: History, Drivers, and Restoration Implications," Annual Report (2018–2019) submitted to South Florida Natural Resources Center, Everglades & Dry Tortugas National Parks, Homestead, Florida, October 2019, 53.

95. Fred H. Sklar, John F. Meeder, Tiffany G. Troxler, Tom Dreschel, Steve E. Davis, and Pablo L. Ruiz, "The Everglades: At the Forefront of Transition," in *Coast and Estuaries*, ed. Eric Wolanski, John W. Day, Michael Elliott, and Ramesh Ramachandran (London: Elsevier, 2019), 277–92.

96. Meeder et al., "Saltwater Encroachment"; Ross et al., "Ecosystem Dynamics in the White Zone," 53.

97. Personal communication, February 18, 2021.

98. Sklar et al., "The Everglades"; Childers, Gaiser, and Ogden, *The Coastal Everglades*.

4. Beyond Resilience

1. Rockefeller Foundation, *The Resilience Age*, directed by Daniel Fries, 2016, https://www.youtube.com/watch?v=w-wDyhewNZ0.

2. Kate Derickson, "Urban Geography III: Anthropocene Urbanism," *Progress in Human Geography* 42, no. 3 (2018): 426.

3. Jayantha Obeysekera, Marjolijn Haasnoot, and Robert Lempert, "How Are Decision-Science Methods Helping Design and Implement Coastal Sea-Level Adaptation Projects?," *US Clivar Variations* 18, no. 3 (2020): 1–7.

4. Alejandro Portes and Alex Stepick, *City on the Edge: The Transformation of Miami* (Oakland: University of California Press, 1993).

5. Kevin Fox Gotham and Miriam Greenberg, *Crisis Cities: Disaster and Redevelopment in New York and New Orleans* (Oxford: Oxford University Press, 2015).

6. Hillary Angelo and David Wachsmuth, "Why Does Everyone

Think Cities Can Save the Planet?," *Urban Studies* 57, no. 11 (2020): 2211.

7. Roger Keil, "An Urban Political Ecology for a World of Cities," *Urban Studies* 57, no. 11 (2020): 2365.

8. See, e.g., Randall W. Parkinson, Peter W. Harlem, and John F. Meeder, "Managing the Anthropocene Marine Transgression to the Year 2100 and Beyond in the State of Florida USA," *Climatic Change* 128, no. 1 (2015): 85–98; Randall W. Parkinson and Danielle E. Ogurcak, "Beach Nourishment Is Not a Sustainable Strategy to Mitigate Climate Change," *Estuarine, Coastal and Shelf Science* 212 (2018): 203–9.

9. Southeast Florida Regional Climate Change Compact Sea Level Rise Ad Hoc Work Group, *Unified Sea Level Rise Projection for Southeast Florida* (2019 update), 4, https://southeastfloridaclimatecompact.org/wp-content/uploads/2020/04/Sea-Level-Rise-Projection-Guidance-Report_FINAL_02212020.pdf.

10. Stephanie Wakefield, *Anthropocene Back Loop: Experimentation in Unsafe Operating Space* (London: Open Humanities Press, 2020).

11. Miles O'Brien, "Will Climate Change Turn Miami into a 'Future Atlantis'?," *PBS Newshour*, October 3, 2018, https://www.pbs.org/newshour/show/will-climate-change-turn-miami-into-a-future-atlantis.

12. Quoted in Jessica Weiss, "Dutch Sea Level Rise Expert: Miami Will Be 'the New Atlantis,' a City in the Sea," *Miami New Times*, May 21, 2015, https://www.miaminewtimes.com/news/dutch-sea-level-rise-expert-miami-will-be-the-new-atlantis-a-city-in-the-sea-7628340.

13. Tom Gustafson, "Islands of South Florida," abstract, University of Florida Water Institute Symposium, Gainesville, Florida, February 6–7, 2018, http://archives.waterinstitute.ufl.edu/symposium2018/downloads/Presentations/Gustafson.pdf.

14. Jeff Goodell, "Miami: How Rising Sea Levels Endanger South Florida," *Rolling Stone*, June 2013, https://www.rollingstone.com/feature/miami-how-rising-sea-levels-endanger-south-florida-200956/.

15. See, e.g., Isaac Stein, "Miami Beach Reimagined" (B. Arch. thesis, University of Miami, 2014).

16. Alliance of the Southern Triangle et al., "Future Cities: Miami Research Intensive Public Session," *The Miami Rail* 23, October 21, 2017, https://miamirail.org/issue-23/future-cities-miami-research-intensive-public-session/; Jeff Goodell, "Miami's Climate Dystopia Gets Real," *Rolling Stone*, July 1, 2021, https://www.rollingstone.com/politics/politics

-features/miami-beach-building-collapse-climate-change-1191989/; Gustafson, "Islands of South Florida"; Gustafson, personal communication, March 4, 2020.

17. Gustafson, "Islands of South Florida," 30.

18. *Prologue: Politics as Palliative Care of the Species*, assembled by Diann Bauer as an Alliance of the Southern Triangle production (2019), 11:21, https://vimeo.com/user43087350.

19. Alliance of the Southern Triangle, "Protocols for the Phase Transition: Towards New Alliances," *Strelka Magazine*, no. 2 (2020).

20. Arch Out Loud, "Miami Floating House Open Ideas Architecture Competition: Results," https://www.archoutloud.com/miami-results.html.

21. Gabriele Filippi Visionary Architectures, "FLO | Floating Housing in Miami: 'Green Water Village,'" https://www.gabrielefilippiarchitetto.com/copia-di-ham-5.

22. Jeffrey E. Huber, Keith Van de Riet, John Sandell, and Lawrence Scarpa, "Salty Urbanism: Towards an Adaptive Coastal Design Framework to Address Sea Level Rise," *The Plan Journal* 2, no. 2 (2017): 389.

23. Brooks + Scarpa, "Salty Urbanism," 2018, https://www.behance.net/gallery/64716915/Salty-Urbanism?locale=en_US.

24. Huber et al., "Salty Urbanism," 405–10; Emily Bloch, "Salt(ier) Life," *Fort Lauderdale Magazine*, July 1, 2019, https://fortlauderdalemagazine.com/saltier-life/.

25. Brooks + Scarpa, "Unplanning Miami: A Sea Level Rise Adaptation Strategy," https://brooksscarpa.com/unplanning-miami; Jeffrey Huber and Michael Horswell, "unPLANningMIAMI: Developing a Transformative Design Framework Plan for Unplanning and Strategic Resettlement of South Florida from the Effects of Sea-Level Rise and Climate Disruption," April 13, 2021, in *Arts and Letters in Conversation*, podcast, season 2, episode 18, 37:19, https://www.buzzsprout.com/471520/8203436-unplanningmiami-developing-a-transformative-design-framework-plan-for-unplanning-and-strategic-resettlement-of-south-florida-from-the-effects-of-sea-level-rise-and-climate-disruption-jeffrey-e-huber-faia-mla-nc.

26. Huber and Horswell, "unPLANningMIAMI."

27. Brooks + Scarpa, "Unplanning Miami."

28. David Chandler, "The Death of Hope? Affirmation in the Anthropocene," *Globalizations* 16, no. 5 (2019): 695–706.

29. Rockefeller Foundation, *The Resilience Age*; Brad Evans and Julian Reid, *Resilient Life: The Art of Living Dangerously* (London: Polity, 2014).

30. Alliance of the Southern Triangle et al., "Future Cities."

31. Kasia Paprocki, "All That Is Solid Melts into the Bay: Anticipatory Ruination and Climate Change Adaptation," *Antipode* 51, no. 1 (2019): 295–315; Kasia Paprocki, *Threatening Dystopias: The Global Politics of Climate Change Adaptation in Bangladesh* (Ithaca: Cornell University Press, 2021).

32. Notably, Fisher prohibited Jews and Black Miamians from renting, buying, or staying in his Miami Beach properties. Ruby Leach Carson, "Forty Years of Miami Beach," *Tequesta: Journal of the Historical Association of Southern Florida* 15 (1955): 3–27.

33. Tyler Mohr, "Dredge," in *Miami Beach: The Project on South Florida*, ed. Rosetta Elkin, working draft Seminar Research Report, Advanced Seminar GSD 09142, Harvard Graduate School of Design, Office for Urbanization, 2015, 45–55, https://issuu.com/harvardofficeforurbanization/docs/mia_seminarbooklet_20160219_fin.

34. Carson, "Forty Years of Miami Beach."

35. Laura A. Ogden and Joel Trexler, "The Everglades as Icon," in *The Coastal Everglades: The Dynamics of Social-Ecological Transformation in the South Florida Landscape*, ed. Daniel L. Childers, Evelyn E. Gaiser, and Laura A. Ogden (New York: Oxford University Press, 2019), 16.

36. Ogden and Trexler, 15.

37. Gustafson, personal communication, March 4, 2020.

38. Jeffrey E. Huber, "unPLANningMIAMI: A Transformative Design Framework Plan for Unplanning and Strategic Resettlement of South Florida from the Effects of Sea Level Rise and Climate Disruption," *Florida/Caribbean Architect*, Winter 2021, 31, https://mydigitalpublication.com/publication/frame.php?i=689836&p=&pn=&ver=html5&view=articleBrowser&article_id=3855163.

39. Neil Brenner, ed., *Implosions/Explosions: Towards a Study of Planetary Urbanization* (Berlin: Jovis, 2014).

40. Stephen Graham, *Cities under Siege: The New Military Urbanism* (London: Verso, 2011), 226.

41. Sahera Bleibleh, Michael Vicente Perez, and Thaira Bleibleh, "Palestinian Refugee Women and the Jenin Refugee Camp: Reflections on Urbicide and the Dilemmas of Home in Exile," *Urban Studies* 56,

no. 14 (2019): 2897–2916; Graham, *Cities under Siege*; Eyal Weizman, *Hollow Land: Israel's Architecture of Occupation* (London: Verso, 2007).

42. Graham, *Cities under Siege*, 84.

43. Michael Roman-John Koscielniak, "Ground Forces: Dirt, Demolition, and the Geography of Decline in Detroit, Michigan" (PhD diss., University of Michigan, 2020).

44. Paprocki, "All That Is Solid"; Paprocki, *Threatening Dystopias*.

45. Paprocki, "All That Is Solid," 296.

46. See, e.g., Parag Khanna, "Climate Migration Is Here. The U.S. Must Invest Accordingly," *Washington Post*, November 17, 2021, https://www.washingtonpost.com/opinions/2021/11/17/climate-change-migration-build-back-better-infrastructure/.

47. Andy Merrifield, *The Politics of the Encounter: Urban Theory and Protest under Planetary Urbanization* (Athens: University of Georgia Press, 2013); Brenner, *Implosions/Explosions*.

48. Katherine McKittrick, "On Plantations, Prisons and a Black Sense of Place," *Social and Cultural Geography* 12, no. 8 (2011): 947–63.

49. Swarnabh Ghosh and Ayan Meer, "Extended Urbanization and the Agrarian Question: Convergences, Divergences and Openings," *Urban Studies* 58, no. 6 (2021): 1097–1119.

50. Martín Arboleda, *Planetary Mine: Territories of Extraction under Late Capitalism* (London: Verso, 2020); Neil Brenner, "Theses on Urbanization," *Public Culture* 25, no. 1 (2013): 85–114; Neil Brenner and Christian Schmid, "The 'Urban Age' in Question," *International Journal of Urban and Regional Research* 38, no. 3 (2014): 731–55; Neil Brenner and Christian Schmid, "Towards a New Epistemology of the Urban?," *City* 19, nos. 2–3 (2015): 151–82; Merrifield, *The Politics of the Encounter*; Lauren Rickards, Brendan Gleeson, Mark Boyle, and Cian O'Callaghan, "Urban Studies after the Age of the City," *Urban Studies* 53, no. 8 (2016): 1523–41.

51. Brenner and Schmid, "The 'Urban Age,'" 750.

52. Brenner, "Theses on Urbanization," 100.

53. Ash Amin and Nigel Thrift, *Seeing Like a City* (Cambridge, UK: Polity, 2016); Bruce Braun, "Environmental Issues: Writing a More-than-Human Urban Geography," *Progress in Human Geography* 29, no. 5 (2005): 635–50.

54. Manuel Castells, *The Rise of the Network Society, The Information Age: Economy, Society and Culture*, vol. 1 (Malden, Mass.: Blackwell, 2009).

55. Thomas Barnett, *The Pentagon's New Map: War and Peace in the Twenty-First Century* (New York: Putnam, 2004), 8.

56. Barnett, 8.

57. Ross Exo Adams, "Natura Urbans, Natura Urbanata: Ecological Urbanism, Circulation, and the Immunization of Nature," *Environment and Planning D: Society and Space* 32, no. 1 (2014): 27.

58. Merrifield, *The Politics of the Encounter*, 6.

59. David J. Madden, "City Becoming World: Nancy, Lefevbre, and the Global-Urban Imagination," in Brenner, *Implosions/Explosions*, 508.

60. Madden, 510; Jean-Luc Nancy, *The Creation of the World, or Globalization* (Albany: SUNY Press, 2007); see also Jean-Luc Nancy and Aurélien Barrau, *What's These Worlds Coming To?*, trans. Travis Holloway and Flor Méchain (New York: Fordham University Press, 2015).

61. Stephen Graham, "Switching Cities Off," *City* 9, no. 2 (2005): 169–94.

62. Gustafson, personal communication, March 4, 2020.

63. Jonathan Pugh and David Chandler, *Anthropocene Islands: Entangled Worlds* (London: University of Westminster Press, 2021).

64. Gustafson, "Islands of South Florida."

65. Gustafson, personal communication, March 4, 2020; also Gustafson, "Islands of South Florida," 30.

66. Brooks + Scarpa, "Unplanning Miami," excerpt from "Framework 3: Transitional and Transformed Economies and Infrastructure," https://brooksscarpa.com/unplanning-miami.

67. Stephen Collier and Andrew Lakoff, "The Vulnerability of Vital Systems: How 'Critical Infrastructure' Became a Security Problem," in *Securing the Homeland: Critical Infrastructure, Risk, and (In)security*, ed. Miriam Anna Dunn and Kristian Søby Kristensen (London: Routledge, 2008), 17–39.

68. Mike Hodson and Simon Marvin, *World Cities and Climate Change: Producing Urban Ecological Security* (London: Open University Press, 2010).

69. Nigel South, "Arcologies, Eco-shelters, and Environmental Exemption: Constructing New Divisions and Inequalities in the Anthropocene," *International Journal for Crime, Justice and Social Democracy* 9, no. 2 (2019): 60–73.

70. Andy Lockhart and Simon Marvin, "Microclimates of Urban Reproduction: The Limits of Automating Environmental Control," *Antipode* 52 (2020): 637–59.

71. Idowu Ajibade, "Can a Future City Enhance Urban Resilience and Sustainability? A Political Ecology Analysis of Eko Atlantic City, Nigeria," *International Journal of Disaster Risk Reduction* 26 (2017): 85–92.

72. See https://www.cityofpraxis.com/.

73. Vincent Callebaut Architectures, *Lilypad*, https://vincent.callebaut.org/object/080523_lilypad/lilypad/projects.

74. Patri Friedman, "SeaSteading: Building on the Platform of the Oceans: Patri Friedman at TEDxSF," April 26, 2012, https://www.youtube.com/watch?v=maLAMwhTjHk; see also Jacob Hurwitz-Goodman and Daniel Keller, *The Seasteaders, dis* (2018), https://dis.art/the-seasteaders; Philip Steinberg, Elizabeth Nyman, and Mauro Caraccioli, "Atlas Swam: Freedom, Capital, and Floating Sovereignties in the Seasteading Vision," *Antipode* 44, no. 4 (2012): 1532–50.

75. Seasteading Institute, "Seasteading! What Did the First Seastead Achieve?," June 6, 2019, https://www.youtube.com/watch?v=ySt3R9DpP4Y.

76. Anthony Cuthbertson, "Elon Musk Reveals Mars 'Acid Test' for Planet to Be Entirely Self-Sufficient," *Independent*, October 20, 2020, https://www.independent.co.uk/space/elon-musk-mars-spacex-starship-colony-b1179088.html.

77. Cuthbertson.

78. Jeff Bezos, "Going to Space to Benefit Earth (Full Event Replay)," *Blue Origin*, May 9, 2019, https://www.youtube.com/watch?v=GQ98hGUe6FM.

79. Iida-Maria Tammi, "Politicians, Pathogens, and Other Threats to Aid Workers: A Material Semiotic Analysis of Violence against Health Care in the Syrian Conflict," *Critical Studies on Security* 9, no. 3 (2021): 190.

80. Brendan Gleeson, *Lifeboat Cities: Making a New World* (Sydney: University of New South Wales Press, 2010); Hodson and Marvin, *World Cities*.

81. Adams, "Natura Urbans."

82. Ajibade, "Can a Future City?"; Sarah Moser, "New Cities: Engineering Social Exclusions," *One Earth* 2 (2020): 125–27.

83. Kate Derickson, "Masters of the Universe," *Environment and Planning D: Society and Space* 36, no. 3 (2018): 556–62; Tariq Jazeel, "Urban Theory with an Outside," *Environment and Planning D: Society and Space* 36, no. 3 (2018): 405–19; Natalie Oswin, "Planetary Urbanization: A View from Outside," *Environment and Planning D: Society and Space* 36, no. 3 (2018): 540–46.

84. Neil Brenner and Roger Keil, "From Global Cities to Globalized Urbanization," *Glocalism: Journal of Culture, Politics and Innovation* 3 (2014): 2, https://glocalismjournal.org/wp-content/uploads/2020/03/brenner-keil_gjcpi_2014_3.pdf.

85. Hillary Angelo and Kian Goh, "Out in Space: Difference and Abstraction in Planetary Urbanization," *International Journal of Urban and Regional Research* 45, no. 4 (2021): 4.

86. Paolo Soleri, *Arcology: The City in the Image of Man* (Cambridge: MIT Press, 1969).

87. Ebenezer Howard, *Garden Cities of Tomorrow* (London: Swan Sonnenschein & Co. Ltd., 1902).

88. Emmanuel Bonnet, Diego Landivar, and Alexandre Monnin, *Héritage et fermeture: Une écologie du démantèlement* (Paris: Éditions Divergences, 2021).

89. James C. Scott, *Seeing Like a State: How Certain Schemes to Improve the Human Condition Have Failed* (New Haven: Yale University Press, 1998).

90. On managed retreat, see, e.g., Orrin H. Pilkey, Linda Pilkey-Jarvis, and Keith C. Pilkey, *Retreat from a Rising Sea: Hard Choices in an Age of Climate Change* (New York: Columbia University Press, 2016). On Manila and Medellín, see Isabelle Anguelovski, Linda Shi, Eric Chu, Daniel Gallagher, Kian Goh, Zachary Lamb, Kara Reeve, and Hannah Teicher, "Equity Impacts of Urban Land Use Planning for Climate Adaptation: Critical Perspectives from the Global North and South," *Journal of Planning Education and Research* 36 (2016): 338. On New York, see Elizabeth Rush, *Rising: Dispatches from the New American Shore* (Minneapolis: Milkweed Editions, 2018). On insurance and securitization in Florida, see Zac J. Taylor, "The Real Estate Risk Fix: Residential Insurance-Linked Securitization in the Florida Metropolism," *Environment and Planning A: Economy and Space* 52, no. 6 (2020): 1131–49.

91. On "enclaving" spatial imaginary, see Morten Nielsen, Jason Sumich, and Bjørn Enge Bertelsen, "Enclaving: Spatial Detachment as an Aesthetics of Imagination in an Urban Sub-Saharan African Context," *Urban Studies* 58, no. 5 (2021): 881–902. On climate migration from Puerto Rico to South Florida, see Jennifer Hinojosa and Edwin Meléndez, "Estimates of Post–Hurricane Maria Exodus from Puerto Rico," Centro: Center for Puerto Rican Studies, Research Brief, Centro RB2017–01, October 2017, https://centropr-archive.hunter.cuny.edu/sites/default/files/RB2017-01-POST-MARIA%20EXODUS_V3.pdf.

92. For their part, Jennifer Rice, Joshua Long, and Anthony Levenda argue that a global "climate apartheid" regime is already emerging via selective infrastructural recalibrations. Jennifer Rice, Joshua Long, and Anthony Levenda, "Against Climate Apartheid: Confronting the Persistent Legacies of Expendability for Climate Justice," *Environment and Planning E: Nature and Space* 5, no. 2 (2022): 625–45.

93. South, "Arcologies," 68.

94. Avi Brisman, Nigel South, and Reese Walters, "Climate Apartheid and Environmental Refugees," in *The Palgrave Handbook of Criminology and the Global South*, ed. Kerry Carrington, Russell Hogg, John Scott, and Máximo Sozzo (Cham: Palgrave Macmillan, 2018), 301–22.

95. Katie Warren, "Fisher Island, the Richest ZIP Code in America...," *Business Insider*, April 14, 2020, https://www.businessinsider.com/richest-zip-code-us-fisher-island-miami-photos-tour-2019-5.

96. Warren.

97. Mike Davis and Daniel Bertrand Monk, *Evil Paradises: Dreamworlds of Neoliberalism* (New York: New Press, 2007).

98. Pugh and Chandler, *Anthropocene Islands*.

99. Timothy Morton, *Dark Ecologies: For a Logic of Future Coexistence* (New York: Columbia University Press): 36.

100. Bruno Latour, *Facing Gaia: Eight Lectures on the New Climatic Regime* (Cambridge, UK: Polity Press, 2017).

101. Shumon Basar, "The Magic Kingdom: How Not to Think about Dubai," *Bidoun*, Summer 2007, https://www.bidoun.org/articles/the-magic-kingdom.

102. Eva Haifa Giraud, *What Comes after Entanglement?* (Durham: Duke University Press, 2019).

103. EcoTech Visions Foundation, *Annual Report 2018*, 3, https://www.etvfoundation.org/copy-of-2021.

104. Invisible Committee, *The Coming Insurrection* (Los Angeles: Semiotext(e), 2009): 58–59; see also Gaston Gordillo, "The Metropolis: The Infrastructure of the Anthropocene," in *Infrastructures, Environment, and Life in the Anthropocene*, ed. Kregg Hetherington (Durham: Duke University Press, 2019), 66–94. For Adams, urbanization must be understood as its own distinct political technology of controlling and organizing space. First developed in nineteenth-century Europe, according to Adams, urbanization is a spatial and political technology of interlinking circulatory flows of people, commodities, and information into a single, continuous system of social, technological, and economic control. For Adams, urbanization is fundamentally a matter of organiz-

ing and controlling space by making and allowing people, capital, commodities, labor to circulate. It is therefore a technology of control via connection and flows, linkage and openness, rather than bounded and demarcated, closed spaces. Ross Exo Adams, *Urbanization and Circulation* (London: SAGE, 2019).

105. Yuk Hui, *The Question concerning Technology in China: An Essay in Cosmotechnics* (Falmouth: Urbanomic Media Ltd., 2016).

106. Friedrich Nietzsche, *Thus Spoke Zarathustra: A Book for None and All*, trans. Walter Kaufmann (New York: Penguin, 1966), 77.

107. Stuart Elden, *The Birth of Territory* (Chicago: Chicago University Press, 2013); Stuart Elden, "Terrain, Politics, History," *Dialogues in Human Geography* 11, no. 2 (2021): 170–89; Phil Steinberg and Kim Peters, "Wet Ontologies, Fluid Spaces: Giving Depth to Volume through Oceanic Thinking," *Environment and Planning D: Society and Space* 33, no. 2 (2015): 247–64; Peter Sloterdijk, *Bubbles: Spheres I* (Los Angeles: Semiotext(e), 2011); Elizabeth Grosz, Nigel Clark, Arun Saldanha, Kathryn Yusoff, Catherine Nash, and Elizabeth Grosz, "Geopower: A Panel on Elizabeth Grosz's *Chaos, Territory, Art: Deleuze and the Framing of the Earth*," *Environment and Planning D: Society and Space* 30, no. 6 (2012): 971–88.

108. Elizabeth Grosz, *Chaos, Territory, Art: Deleuze and the Framing of the Earth* (Durham: Duke University Press, 2008): 69.

109. Huber et al., "Salty Urbanism," 394.

110. Edward W. Soja and Allen J. Scott, "Los Angeles: Capital of the Late Twentieth Century," *Environment and Planning D: Society and Space* 4, no. 3 (1986): 249–54; Edward W. Soja and Allen J. Scott, *The City: Los Angeles and Urban Theory at the End of the Twentieth Century* (Berkeley: University of California Press, 1996).

111. John Law, "What's Wrong with a One-World World?," *Distinktion: Journal of Social Theory* 16, no. 1 (2015): 126–39; Marisol de la Cadena and Mario Blaser, eds., *A World of Many Worlds* (Durham: Duke University Press, 2018).

5. Anthropocene Critical Urban Theory and Practice

1. Sarah Knuth, "All That Is Solid: Climate Change and the Lifetime of Cities," *City* 24, nos. 1–2 (2020): 65–75.

2. Thomas Sugrue, *The Origins of the Urban Crisis: Race and Inequality in Postwar Detroit* (Princeton: Princeton University Press, 2005); Kevin Fox Gotham and Miriam Greenberg, *Crisis Cities: Disas-*

ter and Redevelopment in New York and New Orleans (Oxford: Oxford University Press, 2015).

3. Matthew Gandy, "Learning from Lagos," *New Left Review* 33 (2005): 36–52.

4. David Kilcullen, *Out of the Mountains: The Coming Age of the Urban Guerilla* (Oxford: Oxford University Press, 2013).

5. Jan Nijman, *Miami: Mistress of the Americas* (Philadelphia: University of Pennsylvania Press, 2011), 63; Gregory Bush, "'Playground of the USA': Miami and the Promotion of Spectacle," *Pacific Historical Review* 68, no. 2 (1999): 153–72.

6. Bush, "'Playground,'" 170.

7. Brendan Gleeson, *The Urban Condition* (London: Routledge, 2015): 2; Andy Merrifield, *The Politics of the Encounter: Urban Theory and Protest under Planetary Urbanization* (Athens: University of Georgia Press, 2013), 6. On "the right to the city," see, e.g., David Harvey, "The Right to the City," *New Left Review* 53 (2008): 23–40; Neil Brenner, Peter Marcuse, and Margit Mayer, eds., *Cities for People, Not for Profit: Critical Urban Theory and the Right to the City* (London: Routledge, 2011).

8. Alec Luhn, "Slow-Motion Wrecks: How Thawing Permafrost Is Destroying Arctic Cities," *Guardian*, October 14, 2016, https://www.theguardian.com/cities/2016/oct/14/thawing-permafrost-destroying-arctic-cities-norilsk-russia.

9. Jon Schuppe, "Surfside Collapse Exposes an Overlooked Threat: Saltwater Rising from Underground," *NBC News*, https://www.nbcnews.com/news/us-news/surfside-condo-collapse-salt-groundwater-rcna16473.

10. Jeff Goodell, "Miami's Climate Dystopia Gets Real," *Rolling Stone*, July 1, 2021, https://www.rollingstone.com/politics/politics-features/miami-beach-building-collapse-climate-change-1191989/; Susan Matthews, "The Miami Building Collapse Is a Warning," *Slate*, June 25, 2021, https://slate.com/business/2021/06/miami-building-collapse-climate-change.html; Randall W. Parkinson, "Speculation on the Role of Sea-Level Rise in the Tragic Collapse of the Surfside Condominium (Miami Beach, Florida U.S.A.) Was a Bellwether Moment for Coastal Zone Management Practitioners," *Ocean & Coastal Management* 215, no. 1 (2021): 105968.

11. Alexandre Monnin, "Penser le territoire à l'heure de l'Anthropocène: À propos des 'communs négatifs,'" *AOC*, January 6, 2021, https://aoc

.media/opinion/2021/01/05/penser-le-territoire-a-lheure-de-lanthropo cene-a-propos-des-communs-negatifs/.

12. Lauren Rickards, Brendan Gleeson, Mark Boyle, and Cian O'Callaghan, "Urban Studies after the Age of the City," *Urban Studies* 53, no. 8 (2016): 1523–41.

13. Neil Brenner, "Theses on Urbanization," *Public Culture* 25, no. 1 (2013): 95.

14. David Harvey, "Cities or Urbanization?," in *Implosions/Explosions: Towards a Study of Planetary Urbanization*, ed. Neil Brenner (Berlin: Jovis, 2014), 60.

15. Simin Davoudi, "Imagination and Spatial Imaginaries: A Conceptual Framework," in Simin Davoudi, Ruth Raynor, Bryonie Reid, Bryonie, Jenny Crawford, Olivier Sykes, and David Shaw, "Policy and Practice Spatial Imaginaries: Tyrannies or Transformations?," *Town Planning Review* 89, no. 2 (2018): 99.

16. Davoudi, "Imagination and Spatial Imaginaries," 103.

17. Chiara Bottici and Jamieson Webster, "The Philosophy and Psychoanalysis of the Image: A Conversation," *Public Seminar*, December 10, 2016, https://publicseminar.org/2017/01/the-philosophy-and-psycho analysis-of-the-image/.

18. William Blake, *The Complete Poetry and Prose of William Blake* (Berkeley: University of California Press, 1982), 132.

19. Carol Bacchi, "Why Study Problematizations? Making Politics Visible," *Open Journal of Political Science* 2, no. 1 (2012): 4.

20. Ben Anderson, "What Kind of Thing Is Resilience?," *Politics* 35, no. 1 (2015): 60–66.

21. Kevin Grove, *Resilience* (Abingdon: Routledge, 2018).

22. Michel Foucault, *The History of Sexuality, Volume 2: The Use of Pleasure* (New York: Vintage, 1986), 23.

23. Arkup, "Next Generation Floating House," 2021, https://www.youtube.com/channel/UCZRHwjSJ5qS7inq5haXnl4Q.

24. David Farrier, *Anthropocene Poetics: Deep Time, Sacrifice Zones, and Extinction* (Minneapolis: University of Minnesota Press, 2019). Frédéric Neyrat, *The Unconstructable Earth: An Ecology of Separation* (New York: Fordham University Press, 2019).

25. Timothy Morton, *Dark Ecology: For a Logic of Future Coexistence* (New York: Columbia University Press, 2016).

26. Anna Tsing, *The Mushroom at the End of the World* (Princeton: Princeton University Press, 2016), 8.

27. David Chandler, *Ontopolitics in the Anthropocene: An Introduction to Mapping, Sensing and Hacking* (London: Routledge, 2018).

28. Olivier Faye, "Newly Pessimistic Macron Presents Hat-Trick of Doom," August 25, 2022, *Le Monde*, https://www.lemonde.fr/en/politics/article/2022/08/25/newly-pessimistic-macron-presents-a-hat-trick-of-doom_5994680_5.html.

29. Clive Barnett, *The Priority of Injustice: Locating Democracy in Critical Theory* (Athens: University of Georgia Press, 2017); Stephen Collier, "Topologies of Power: Foucault's Analysis of Political Government beyond 'Governmentality,'" *Theory, Culture & Society* 26, no. 6 (2009): 78–108.

30. Chandler, *Ontopolitics in the Anthropocene*, xvi.

31. Arkup, "Next Generation Floating House."

32. Critiques of austerity and the small scale are not new. From a socialist perspective, Matt Huber's book *Climate Change as Class War* advocates against climate-change-driven neoliberal austerity with a politics of "more" through more low-carbon jobs, housing, etc. Matt Huber, *Climate Change as Class War* (London: Verso, 2022). Or we might think of Leigh Phillips's *Austerity Ecology and the Collapse-Porn Addicts*, which is also a socialist, pro-planning, pro-nuclear takedown of small is beautiful and eco-commune politics. Leigh Phillips, *Austerity Ecology and the Collapse-Porn Addicts: A Defence of Growth, Progress, Industry, and Stuff* (Winchester, UK: Zero Books, 2015). But, in contrast, for Anthropocene critical urban theory/practice, the goal is not necessarily to reject one approach—climate austerity or small is beautiful—with a single appropriate solution (e.g., a Green New Deal policy package, total world socialist rational society, etc.). Among other criticisms one could make of such an approach, it would immediately close down the exploration and experimentation the approach seeks to open.

33. Alliance of the Southern Triange, "Protocols for the Phase Transition: Towards New Alliances," *Strelka Magazine*, no. 2 (2020).

34. Nigel Clark, *Inhuman Nature: Sociable Life on a Dynamic Planet* (Los Angeles: Sage, 2011); Nigel Clark, "'Primordial Wounds': Resilience, Trauma, and the Rifted Body of the Earth," in *Resilience in the Anthropocene: Governance and Politics at the End of the World*, ed. David Chandler, Kevin Grove, and Stephanie Wakfield (Abingdon: Routledge, 2020), 147–61.

35. Nigel Clark, "Enflamed Imaginations: Of Fire and Futurity," in *Culture and Climate Change: Scenarios*, ed. Renata Tyszczuk, Joe Smith,

and Robert Butler (Cambridge: Shed, 2019), 30; Nigel Clark, "Vertical Fire: For a Pyropolitics of the Subsurface," *Geoforum*, April 2020; Nigel Clark and Kathryn Yusoff, "Queer Fire: Ecology, Combustion and Pyrosexual Desire," *Feminist Review* 11, no. 1 (2018): 7–24.

36. Clark, "Enflamed Imaginations," 30.

37. See, e.g., Kimberley Peters, Philip Steinberg, and Elaine Stratford, eds., *Territory beyond Terra* (London: Rowman and Littlefield, 2018).

38. Julian Reid, "'We the Resilient': Colonizing Indigeneity in the Era of Trump," *Resilience: International Policies, Practices and Discourses* 7, no. 3 (2019): 255–70.

39. Clark and Yusoff, "Queer Fire."

40. Yuk Hui, *The Question concerning Technology in China: An Essay in Cosmotechnics* (Falmouth, UK: Urbanomic Media Ltd., 2016).

41. Yuk Hui, "Cosmotechnics as Cosmopolitics," *e-flux*, no. 86 (November 2017), https://www.e-flux.com/journal/86/161887/cosmotechnics-as-cosmopolitics/.

42. Yuk Hui, "Rethinking Technodiversity," *The UNESCO Courier*, March 31, 2023, https://courier.unesco.org/en/articles/rethinking-technodiversity#:~:text=The%20noodiversity%20has%20been%20maintained,in%20exchange%20with%20other%20localities.

43. Brent Weisman, *Unconquered People: Florida's Seminole and Miccosukee Indians* (Gainesville: University of Florida Press, 1999).

44. Houston Cypress, personal communication, February 26, 2023; Houston Cypress, "Miccosukee Native Houston Cypress," *Tomorrow's Ancestors Speak*, 2014, https://www.youtube.com/watch?v=sG50Ux0txKI; Dara Friedman, interview with Houston Cypress, *Miami Rail*, Winter 2016, https://miamirail.org/winter-2016/interview-with-houston-cypress/.

45. Personal communication, February 26, 2023.

46. See, e.g., Terence Cantarella, "Meet the Owner of Miami's Last Floating Home on Biscayne Bay," *Miami New Times*, December 7, 2015, https://www.miaminewtimes.com/news/meet-the-owner-of-miamis-last-floating-home-on-biscayne-bay-8098501; "Flashback Miami: Houseboats," *Miami Herald*, 2015, https://flashbackmiami.com/2015/06/10/houseboats/#lightbox[group-6299]/34; *Stiltsville: Generations on the Flats*, WLRN Documentaries, 2010, https://www.pbs.org/video/wlrn-documentaries-stiltsville-generations-flats/?_ga-ft=ZbQG7Q.AA.AA.AA.AA.k03-0z3RSHuSHRxqNUTsjg..0.

47. On this see Irvin D. S. Winsboro and Joe Knetsch, "Florida

Slaves, the 'Saltwater Railroad' to the Bahamas, and Anglo-American Diplomacy," *Journal of Southern History* 79, no. 1 (2013): 51–78.

48. Germane Barnes, "Black Miami's Resiliency: A Photographic Essay," *Anthurium: A Caribbean Studies Journal* 16, no. 1 (2020): 3.

49. Laura Ogden, *Swamplife: People, Gators, and Mangroves Entangled in the Everglades* (Minneapolis: University of Minnesota Press, 2011), 21–22.

50. Quoted in Christopher Flavelle, "Miami Will Be Underwater Soon. Its Drinking Water Could Go First," *Bloomberg*, August 29, 2018, https://www.bloomberg.com/news/features/2018-08-29/miamis-other-water-problem.

51. Gillian Rose, *Feminism and Geography: The Limits of Geographical Knowledge* (Cambridge, UK: Polity Press, 1993).

52. Tsing, *The Mushroom*.

53. Jasper Bernes, "Planning and Anarchy," *South Atlantic Quarterly* 111, no. 1 (2020): 68–69.

54. Brad Evans and Julian Reid, *Resilient Life: The Art of Living Dangerously* (London: Polity, 2014).

55. Clark, "Enflamed Imaginations," 7.

56. Mike Davis and Daniel Bertrand Monk, *Evil Paradises: Dreamworlds of Neoliberalism* (New York: New Press, 2007).

57. On subaltern resistances to urban climate projects, see Vanesa Castán Broto, Enora Robin, and Aidan While, *Climate Urbanism: Toward a Critical Research Agenda* (Cham: Palgrave Macmillan, 2020). On "zombie" systems, see Monnin, "Penser la territoire."

58. Monnin, "Penser la territoire"; Emmanuel Bonnet, Diego Landivar, and Alexandre Monnin, *Héritage et fermeture: Une écologie du démantèlement* (Paris: Editions Divergences, 2021).

59. Giorgio Agamben, *The Use of Bodies* (Stanford: Stanford University Press, 2015), 273.

60. Emmanuel Bonnet, Diégo Landivar, Alexandre Monnin, and Laurence Allard, "Le design, une cosmologie sans monde face à l'Anthropocène," *Sciences du Design* 2, no. 10 (2019): 99.

61. Bonnet et al., 101.

62. See, e.g., Elinor Ostrom, *Governing the Commons: The Evolution of Institutions for Collective Action* (Cambridge, UK: Cambridge University Press, 1990); Michael Hardt and Antonio Negri, *Common Wealth* (Cambridge, Mass.: Belknap Press of Harvard University Press, 2009).

63. Tsing, *The Mushroom*; Donna Haraway, *Staying with the Trou-*

ble: Making Kin in the Chthulucene (Durham: Duke University Press, 2016).

64. Alexandre Monnin, "It Must Happen in France," *ARQ (Santiago)*, no. 111 (2002): 79.

65. Bonnet et al., "Le design," 101.

66. Bonnet et al., 101.

67. Bonnet et al., 101.

68. Emmanuel Bonnet, Diégo Landivar, and Alexandre Monnin, "Crise climatique: 'Nous devons apprendre à désinnover,'" *Le Monde*, September 3, 2021, https://www.lemonde.fr/idees/article/2021/09/03/crise-climatique-nous-devons-apprendre-a-desinnover_6093287_3232.html.

69. Gean Moreno, "Glitch-People," *Affidavit*, November 21, 2016.

70. David Graeber and David Wengrow, *The Dawn of Everything: A New History of Humanity* (New York: Farrar, Straus and Giroux, 2021).

71. Graeber and Wengrow, 283, emphasis added.

Coda

1. Domingo Castillo, *Tropical Malaise: Prologue*, 2017, HD film, context-contingent screening, 18:22. Directed and edited by Domingo Castillo, music by Nick Klein, narration by Annie Bandez, sound by Andrew Yeomanson, Barbara Elting, James Mckillop, audio mastering by Jack Callahan. *Tropical Malaise: Prologue* was commissioned by SEA LEVEL RISE, a series of site-specific temporary public art interventions exploring the topic of rising seas and its impact in South Florida organized by Art in Public Places of Miami-Dade County Department of Cultural Affairs and the University of Miami School of Communication. Additional support was provided by BFI Miami, Obsolete Media Miami, Art Center South Florida, City of Miami Beach Cultural Affairs Office, and Primitive Languages.

2. Natalia Zuluaga, "Point and Pivot," in *Between the Material and the Possible: Infrastructural Re-examination and Speculation in Art*, ed. Bassam El Baroni (Cambridge: MIT/Sternberg, 2022), 132.

3. Gean Moreno, "Glitch-People," *Affidavit*, November 21, 2016.

4. Zuluaga, "Point and Pivot," 133.

5. Jefferson Cowie and Joseph Heathcott, "The Meanings of Deindustrialization," in *Beyond the Ruins: The Meanings of Deindustrialization*, ed. Jefferson Cowie and Joseph Heathcott (Ithaca, N.Y.: ILR Press, 2003), 4.

6. Cowie and Heathcott, 5.

INDEX

AAA, 67
Aalbers, Manuel, 98
Action Plan, 170
Adams, Ross Exo, 109, 153, 163, 254n84, 263n104
ADaPT: Adaptation Design and Planning Tool for Urban Areas in the Coastal Zone, 142
adaptation, 141, 174, 175, 183, 197; capacity for, 98, 176, 182; coastal, 135, 145; cycle, 69 (fig.); designs, 32, 39; environmental, 173; extreme, 137–39; human-environment, 7, 185, 186; reactive, 190. See also climate adaptation
Addie, Jean-Paul, 102
Adey, Peter, 76
Agamben, Giorgio, 3, 194
agglomerations, 6, 11, 18, 38, 123, 145, 152, 153, 197
agriculture, 96; commercial, 43; vertical, 140
Akon City, 157
Alexander, Christopher, 121
algae blooms, 118

Alliance of the Southern Triangle (A.S.T.), 140, 141, 155, 158, 183
Allman, T. D., 33
Amazon, 60
Amin, Ash, 90, 101–2
Anderson, Ben, 76, 177, 216n61
Angelo, Hillary, 32, 119, 137–38, 159
Antarctic ice melt, increase in, 93
Anthropocene: characterization of, 4; defining, 2, 3; hegemonic understandings of, 108; human life in, 198; impact of, 3, 52; living in, 92–93, 180; modalities of, 36; perpetuation of, 7; responding to, 178; term, 6; upheavals of, 5; uncertainty of, 9; urban, 4, 27
Anthropocene critical urban theory/practice, 5–9, 10, 19–20, 21, 24, 35, 41, 174–75, 178, 183, 185, 186, 196–97; need for, 39; research pathways for, 195

| 271

INDEX

Anthropocene: The Human Epoch (film), 154
aquaculture, 8, 150, 157
Aradau, Claudia, 70, 89, 91
Arboleda, Martín, 7
Arch Out Loud, 141, 142
architecture, 20, 189; amphibious, 107; Black, 188; storm-resistant, 155
Arctic ice melt, 48
Arendt, Hannah, 40
Arkup, 180, 182; livable yacht, 179, 180 (fig.)
Armstrong, Louis, 95
art: Anthropocenic, 193; climate-change-themed, 49; exhibitions, 17, 81
Art Basel fair, 81
Art of the Long View: Planning for the Future in an Uncertain World, The (Schwartz), 81
Ashley, Bill, 188–89
Aspen Ideas: Climate Conference, 50
assimilation, 101, 186, 187
A.S.T. See Alliance of the Southern Triangle
Atlantic Ocean, 10, 11, 50; advance of, 109–10
Atlantis, 1, 49, 61, 77, 138, 201
austerity, 105, 169; climate, 161, 267n32; economic, 181; green, 151, 181
Avengers: End Game (film), 199

Bacchi, Carol, 177
Bacigalupi, Paolo, 190
Back Bay Coastal Storm Risk Management Study, 119
back loop, 67–73, 69 (fig.), 79, 181
Bahamas, 10, 188
Ballard, J. G., 169
Bankers Insurance, 67
Barnes, Germane, 189
Barnett, Allain, 22, 39, 240n25
Barnett, Clive, 19, 20, 181
Barnett, Thomas, 153
barriers, 14, 27, 50, 88, 151
Basar, Shumon, 163
Bauer, Diann, 140
Before the Flood (DiCaprio), 54
Bennett, Jane, 127
Bernardo, Melissa, 105–6
Bernes, Jasper, 191
Bertelsen, Bjorn Enge, 102–3
Bezos, Jeff, 60, 157
Big Data, 30, 32
Bigger, Patrick, 120
Billionaire's Bunker, 60
Billionaires' Row, 8
biodiversity, 3, 68, 75, 120, 186
biopolitics, 52, 98, 107, 123, 127, 131, 179
bioswales, 143
Biscayne Aquifer, 109, 110 (fig.), 124
Biscayne Bay, 42, 50, 74, 75, 119, 124, 141; dredging, 146; flooding in, 32
Bitcoin, 59
Black, Jack, 55
Blacks, 94–95, 258n32
Blackstone, 60
Blake, William, 177
Blockchain.com, 60
Bloomberg, 63
Blue Frontiers, 157

Bonnet, Emmanuel, 160, 194
Bosh, Chris, 74
Bottici, Chiara, 28, 84, 176
Boyer, M. Christine, 90
Boyle, Mark, 40
Brassett, James, 76
Bratton, Benjamin, 9, 32
Braun, Bruce, 117, 123, 127
Bravo, Thomas, 60
breakwaters, 143
Brenner, Neil, 5, 40, 166, 174; operational landscape and, 116; urban theory and, 6, 10; urbanization and, 18, 159
Briceno, Henry, 138
Bridge, Gary, 19
Brookings Institution, 167
Brooks + Scarpa, 142, 143, 147, 155, 158, 167, 194
Broto, Vanesa Castán, 52, 53
Bruce Mowry Day, 75
buffer zone, 68
Bulkeley, Harriet, 52, 53, 56, 71–72
Bush, Gregory, 43, 170
Bush, Jeb, 114
bushcraft, 106
Business Case for Resilience in Southeast Florida, The (Urban Land Institute), 61
Business Insider, 44

Cajun Navy, 105
California Forever, 157
Callahan, Jack, 270n1
Callebaut, Vincent, 157
Calusa, 113, 147
Caminos, Ximena, 48
Cape Hatteras, 12

capitalism, 6, 7; global, 5, 166
Capitalocene, 4, 6
carbon sequestration, 53
Carse, Ashley, 116
Cason, James, 115
Castillo, Domingo, 199, 200–201, 270n1
Castoriadis, Cornelius, 28
Castro, Fidel, 81
Cattelino, Jessica, 37, 128
Cava, Daniella Levine, 84
Center for Resilient Studies and, Landscapes, 100 Resilient Cities' Resilience Accelerator program of, 50
CERP. See Comprehensive Everglades Restoration Plan
Champlain Towers South, collapse of, 172
Chandler, David, 155, 180–81, 182
Chattopadhyay, Swati, 26, 117
Chokoloskee Island, 146
CI Financial, 60
Citadel Securities, 60
"Cities or Urbanization?" (Harvey), 169
Citizens, 67
city: artificial, 121; at end, 10–15, 169–82; global, 201; hero futurist, 157; infrastructural, 102; lifeboat, 158; nature and, 37, 129–33; rational, 90; zombie, 193
City of Miami, 15, 94
City of Miami Beach, 15, 50, 202, 213n33, 240n28
CityCoin, 58
civilization, 3, 26; escaping, 157;

rebuilding, 106–7; starter kits, 106
Clark, Nigel, 183, 191, 192
class: exploitation of, 197; realities of, 196
climate adaptation, 9, 42, 50, 65, 66, 149, 151, 161, 174; extreme, 162; infrastructures of, 34; urban, 7, 8, 166–67, 171
climate change, 16, 18, 19, 35, 39, 53, 55, 61, 70, 72, 79, 82, 85, 106, 107, 117, 122, 123, 141, 144, 161; attitude toward, 64; challenges of, 41–42, 86; denying, 139; endgame of, 109; first responders to, 14; global, 32; governing, 120; impact of, 1, 3, 7, 12, 23, 32, 44, 49, 52, 54, 62, 173, 175; increasing, 133; political-economic landscape and, 12; responses to, 5; risks in, 130; solution for, 138; testing solutions to, 52; threats from, 15, 65; urban, 112, 171; vulnerability to, 86; withstanding, 126
Climate Change Advisory Task Force, 15
climate change ground zero, 33, 35, 44, 49, 56, 57–58, 61, 62, 64, 65, 77, 201
climate insurance bubble, 66–67
climate policy, Indigenous knowledge in, 129
climate refugees, ecopolis for, 157
climate risk, 44, 71, 120
cocaine trafficking, 170
Coconut Grove, 188

Collier, Stephen, 181
Collins, Phil, 1
colonialism, 31, 89, 129
Colorado River, 190
Colored Town, 94
Columbia University, 50
"command and control" solution, 86, 113, 120
commodity chains, mapping, 24
communication, 6, 27, 97, 186
Community Emergency Response Team (CERT), 101
Comprehensive Everglades Restoration Plan (CERP), 111, 114, 126, 254n77; conceptual plan, 131; founding of, 115; projects by, 132
"Confession of the Flesh, The" (Foucault), 20
Connolly, N. D. B., 26, 95
conservation: biological, 120; shift away from, 120; water, 125
Coolest Block Contest, 72
Coral Gables, 115
cosmotechnics, 164, 185–86, 192, 195
Covid-19 pandemic, 36, 60, 106, 156, 162; impact of, 57–58, 82; islandizing and, 157
Cowie, Jefferson, 202
Cox, Savannah, 39, 63–64, 66, 240n25; resilience building and, 22
credit ratings, 62, 65
crime rates, 95
crises: community response to, 101–2; environmental, 7, 123, 179; naturalization of, 98; permanent, 179; urban, 131

critical infrastructure, 24, 26, 35, 99, 102, 117, 124, 175, 176; biopolitical role of, 123; Everglades as, 126, 128; nature as, 123
Crutzen, Paul, 3
cryptocurrency, 1, 11, 36, 58, 75, 78, 157, 196
culture, 165, 192; Anthropocenic, 189, 193; civic, 86; Miccosukee, 187; popular, 44
culture wars, 1
Curtis + Rogers Design Studio, visualization by, 119 (fig.)
Cypress, Houston, 187

Dalby, Simon, 52, 76
Damon, Matt, 74
Daniels, Stephen, 29
data visualization, 30, 31
Davis, Mike, 162, 192
Davis, Sammy, Jr., 95
Davoudi, Simin, 29, 30, 175
De Young, Raymond, 54
degrowth, notion of, 195
deindustrialization, 169, 202
delinking, 35, 154, 155, 156, 157, 158, 162, 163–64, 165–66, 173
demolition, 147, 150, 151, 173
Department of Homeland Security, 100
Department of Unplanning, 147, 148
Derickson, Kate, 136
desalination, 155, 157, 179, 190
desaturation, 195
destruction: ecological, 150–51; environmental, 52, 114, 131; social, 132, 150–51
Detroit, 135

Detroit Demolition Program, 150
development. See urban development
DiCaprio, Leonard, 54
Didion, Joan, 33, 43, 151
dikes, 14, 126
disasters, 16, 36, 93, 94, 96, 97, 102, 103, 104, 145, 156, 179, 193; management, 70, 120; natural, 18, 85, 87; naturalization of, 98; social media and, 100; surviving, 107
dispositifs, 19, 21, 31, 37, 70, 110, 116, 181
doomsday bunkers, 156
Douglas, Marjory Stoneman, 118
Downtown Miami, potential submergence for, 14
Drowned World, The (Ballard), 169
drug cartels, 11
Dunn, Marvin, 95
Dust Bowl, 110
Dutton, Andrea, 12

East River Park, bulldozing of, 8
East Side Coastal Resiliency Project, 8
Easterling, Keller, 25
Echemendia-Kouri, Jean, 74
eco-austerity regimes, 181
ecodistricts, 156
ecological crisis, 2, 72, 73, 78, 137
ecological infrastructure (EI), term, 120
ecology, 8, 22; island, 193; resilience, 70; transformed, 156
economic conditions, 73, 145

economic growth, 26, 63
economic reputation, 62, 66
economic systems, 70, 71, 73
ecopolis, 157
ecopreneurship, 163
ecosystems, 126, 130, 133, 187; aquatic, 80; degradation of, 114; restoration of, 115, 120; submergence of, 132; sustainability of, 158
EcoTech Visions Foundation, 106, 163
Edwards, Gareth, 71–72
Eko Atlantic City, 156, 166
Elden, Stuart, 165
elevated roads, 51, 55, 65, 71, 72, 118, 149
Elliott-Ortega, Kara, 117
emergency preparedness initiatives, 100–101
end: city at, 169–82; cosmotechnics at/for, 182–86
endangered species, 114
energy, 14, 38, 148, 186
entropocene, 185–86
environment: Anthropocenic, 190; built, 25; urban, 27, 193
Envisioning the Future Urban Resilience to Climate Change in Greater Miami Area (URExSRN), 87–88
epistemologies, 21, 197; objectivist, 92; salvage-oriented, 183
epoch of the humans, urban resilience and, 173
Erlich, Leandro: sculptures by, 48 (fig.)
ethos, 21
Evans, Brad, 103
Evans, James, 53, 54, 73

Everglades, 10, 11, 26, 110, 115, 120, 122, 124, 146–47, 186, 187, 188–89; backcountry, 111 (fig.); bureaucratization of, 128; as carbon emitter, 133; city and, 129; flooding, 188; highway through, 125–26; hydrologic flows of, 112 (fig.); as infrastructure, 125, 126, 128, 131, 133, 149, 253n77; liquid landscape of, 116; management of, 114, 117, 118, 187; operationalizations of, 152; reclaimed land in, 114 (fig.); reimagining, 116, 128; restoration of, 114, 132; saltwater intrusion and, 136; sea-level rise and, 132; urban government and, 112; water flows from, 111, 113, 254n77; as water management system, 253n77; wetland ecosystem, 37
Everglades National Park, 124
Everglades Skyway bridge system, 126
experimentation, 6, 18, 68, 92, 113, 167, 203; back loop, 79; climate, 36, 77; conceptual, 174; elite, 183; governing by, 52; imaginal, 136; infrastructural, 71, 136; methodological, 174; practical, 40, 174, 175; small, 54; spaces of, 19; theaters of, 31; urban, 35, 52–56, 71, 73, 86, 173, 193, 195–98
extrastatecraft, 25

Faena Hotel, 81
Farmers Insurance, 67

INDEX | 277

FarmTech, 106
Farrier, David, 179
Fast Company (film), 46
Federal Housing Administration, 26
feedback, 30, 92, 139, 176
FEMA, 100
Filippi, Gabriele, 141, 160
finance, 58, 67, 122
Financial Times, 58
First Street Foundation, 66–67
Fisher, Carl, 16, 146, 161, 258n32; publicity stunts by, 45 (fig.)
Fisher Island, 161
Flagler railroad, 94
floating cities/villages, 141, 156, 157, 160
flooding, 32, 46, 66, 71, 85, 188; coastal, 51; groundwater, 51; high-tide, 51; impact of, 66; managing, 14, 88, 127, 138, 148; projections on, 13–14; sea-level rise and, 12, 22, 44, 62; storm-related, 51; tidal, 61; urban, 65
Florida Atlantic University, 142
Florida Coastal Everglades Long Term Ecological Research Project, 37
Florida Department of Transportation, 125
Florida Disaster Relief Systems, 97
Florida House of Representatives, 138
Florida International University, 64, 138
Florida Keys, 139, 190; sea-level rise and, 46

flows, 111, 113; Anthropocene, 130; flows against, 124–29; Holocene, 130
food, 186, 188; growing, 156; production, 106, 148; shortages, 96
Fort Lauderdale, 142, 144
fossil fuels, 117
Foucault, Michel, 23, 25, 178; dispositif and, 19, 21, 31; on liberalism, 127; problematizations and, 20, 21
freshwater-saltwater interface, 125
Friedman, Milton, 157
front loop, 67, 68, 69 (fig.)
FTX, collapse of, 59
Fukushima nuclear accident, 135, 193
Future of the American City Design Studio, 240n28
futures scenarios, 94, 104, 240n28; conservative nature of, 88–89; problem of, 87–88

Gabriele Filippi Architetto, work of, 142 (fig.)
Gabrys, Jennifer, 32, 79
Gaia, 185
Gandy, Matthew, 116, 130, 170
Garden City, 159
gardens: high-tide, 143; rain, 143
Gehry, Frank, 199
Gelber, Dan, 51, 56, 63, 65, 78
general obligation bonds, 65
geology, 3, 22, 44, 167
geopolitics, 37
geopower, 127
Getting Ahead of the Curve for a Resilient Economy: An

278 | INDEX

Introduction to Risks and Opportunities of Climate Change for Top Business Sectors in Southeast Florida, 61
GHGs, 138–39
Ghosh, Swarnabh, 6
Gieseking, Jen Jack, 93
Giraud, Eva Haifa, 163
Gladesmen: Gator Hunters, Moonshiners, and Skiffers (Ogden), 188
Gleeson, Brendan, 40, 135, 158, 172
Global Health Initiatives, 93
global warming, 15, 70, 128
Goh, Kian, 7, 8, 159
González, Elián, 1
"Goodbye Miami" projections, 14–15
Goodell, Jeff, 49
goTenna Mesh devices, 105
governance: adaptive, 52; anticipatory, 70, 71; biopolitical, 123; climate change and, 3, 120; eco-, 127; global, 84; modes of, 70; neoliberal, 153; risk, 76; techniques, 133; technologies of, 137. See also urban governance
Graeber, David, 197
Graham, Bob, 125
Graham, Stephen, 150, 154
Grand Theft Auto: Vice City (film), 44
Great Land Hustle, The (Paulson), 78
Great Recession, 57
Great Wall, 191
Greater Everglades, 132
Greater Miami, 213n33; adaptations in, 10; pandemic and, 60; tourism and, 61
Greater Miami and the Beaches, 50, 84, 240n28; creation of, 15
Green Jacket, The, 143
Green New Deal, 267n32
Green Water Village, 141, 154, 160
"Green Water Village" (Filippi), 141, 142 (fig.)
Greenland ice melt, 48
Gregory, Derek, 92
Griffin, Ken, 60
Grodin, Felice, 140
Grosz, Elizabeth, 166
groundwater, rising, 148
Grove, Kevin, 39, 86, 107, 240n25; resilience building and, 22
Guardian, 44, 46
Gunderson, Lance, 118
Gustafson, Tom: GHGs and, 138–39; Islands of South Florida and, 139, 154, 158; vision of, 147

Hadid, Zaha, 109
Halpern, Orit, 18
Harlem, Peter: maps by, 46, 47 (fig.), 49
Harris, Kamala, 50
Harvard Graduate School of Design, 15, 240n28
Harvey, David, 169, 174
heat waves, 2, 96
Heathcott, Joseph, 202
Hernández, Patricia Margarita, 140
heterotopias, 72
Heur, Basvan, 54

Hibiscus Island, 51, 146
Hill-Gabriel, Julie, 125
Hirst, Damien, 81
Hodson, Mike, 158
Holiday, Billie, 95
Holling, C. S., 16, 17, 118
Holocene, 2, 3, 129, 164, 203
Homestead, 96
houseboats, 188
housing, 148; affordable, 103, 141; floating, 188; low-carbon, 53; lower-income, 84; market, 84, 98, 106; prices, 57, 58, 82, 103, 141; public, 26
Howard, Ebenezer, 159, 160
Huber, Jeff, 142, 144, 147, 167
Hui, Yuk, 164, 185, 186
human agency, 28, 103, 184, 191, 192
Human Epoch, 6, 40
Hurricane Andrew, 96
Hurricane Irma, 96
Hurricane Katrina, 118
Hurricane Maria, 94, 161
Hurricane Sandy, 16, 93, 100
hurricanes, 14, 15, 16, 85, 94, 96, 98, 100, 106, 107, 118, 138, 161; Category 5, 93; exposure to, 149
Hydro Authority, 88
hydrology, 125, 128

imaginaries, 40, 62, 76, 98, 103, 149, 171, 191, 197; adaptation, 170, 177; Anthropocenic, 34, 170, 182, 199–200; apocalyptic, 170; climate, 151, 177, 179; climate change, 31, 32, 65, 78; creating/ internalizing, 28–29; dominant, 175, 176, 179, 186; emergence of, 33; end-of-the-city, 172; Everglades, 113; exploring, 32–33; geographical, 29, 30; governing, 104, 175, 185; imagination and, 28, 33; infrastructure and, 35, 42, 103, 126, 176; island, 39, 126, 140–45, 149, 150, 151, 152, 154, 162, 164, 165, 166, 171; media, 41, 136, 171; post-pandemic, 56; resilience, 34, 39, 76, 137, 140, 165, 166, 176, 181, 186; sea-rise, 39–40; sociotechnical, 32; spatial, 28–35, 89, 92, 161; transformative, 88; understanding, 185; urban, 39, 57, 65, 104, 137, 195; urbicidal, 38
imagination, 8, 9, 24–35, 43, 104, 152, 174, 192, 196, 203; climate change, 19; free, 34, 175; human, 182; imaginaries and, 28, 33; individual, 29; political, 25, 84, 104; productive, 176; reappropriation of, 182–83; regime of, 103; resilience, 22, 23, 177; revalorization of, 84; social, 104; spatial imaginaries and, 28–35; urban, 22, 23, 162
"In the Air Tonight" (Collins), 1
Indian Creek, 60, 67, 75
Indian Wars, 116, 187
Indigenous people, 111; ecosystem degradation and, 114
industrial liberal society, 41, 175
inequality, 158; economic, 2, 22, 44, 82; social, 85; socioeconomic, 7, 84, 86, 240n25

infinity loop, 72
infrastructure, 8–9, 23, 24–28, 34, 53, 72, 94, 117, 122, 128, 143, 155; aging, 65, 150, 152; Anthropocenic, 18, 41, 136, 189; biopolitical, 131; blue, 118, 119; building, 108, 179; communication, 6; decommissioning of, 148; definitions of, 22; ecological, 37, 120, 137, 145; Everglades as, 126, 131, 133, 149, 253n77; experimental, 113; food production, 148; global, 153, 156; gray, 53, 88, 118, 124, 136; green, 88, 118, 119, 136; human, 37, 100; industrial, 137, 158; Lidar mapping of, 15; living, 8, 53, 148; local, 106, 158; maritime, 188; modern, 117, 129; mutations of, 4–5; nature as, 120, 122, 123, 127, 129, 130; neoliberal, 153, 192; people as, 36, 99, 101; political, 117; projects, 67, 131; recalibrating, 99–100; resilience, 15, 18, 22, 23, 37, 50, 63, 66, 71, 129, 136, 137, 138; sea-rise, 39–40; self-organizing, 91; stormwater management, 50; technical, 27, 37, 113, 128, 137, 145; urban life and, 26; water management, 26. See also critical infrastructure; social infrastructure; urban infrastructure
Ingels, Bjarke, 14, 91
Institute of Environment (FIU), 138
insurance crisis, 66–67

interlinking, 35, 38, 137, 153, 158, 163, 167, 173, 175, 264n104
International Panel on Climate Change, 12
International Space Station, photo from, 11 (fig.)
International Water Affairs of the Netherlands, 55
Invisible Committee, 163
island territories, 151, 166
islandization, 38, 39, 40, 140, 142, 143, 156, 158, 159, 160–61, 162, 166; post-resilience, 171; rise of, 157; strategic, 157
islands, 179; artificial, 146, 147, 155; economic, 158; floating, 141, 154; infrastructural, 158; political, 158
Islands of South Florida, 141, 145, 146, 149, 154, 158, 163, 164, 165, 166; building, 139; vision of, 140
Israeli Defense Forces, 149
Iwaniec, David, 89, 105

Jacobs, Jane, 91, 121, 241n40
Jenin camps, 149
Jim Crow, 11, 44, 82, 95

Kahn, Herman, 85
Kaika, Maria, 130
Kallianos, Yannis, 25
Kaplan, Stephen, 54
Karvonen, Andrew, 54
Kedan, Elite, 140
Keil, Roger, 6, 10, 138, 159
Key Biscayne, sea-level rise and, 46

Key West, sea-level rise and, 49
King Tides, 12, 51, 54; staircase at, 13 (fig.)
Kissimmee River, 124
Klinenberg, Eric, 100
Knuth, Sarah, 169–70, 172
Koopman, Colin, 21
Koscielniak, John, 150

Lagos, 9, 17, 157, 170
Lake Okeechobee, 124
Land Adjust (Islands and Atolls), 143; scenario from, 144 (fig.)
land reclamation, 140, 165
Landivar, Diego, 160, 194
Larkin, Brian, 25, 26, 34
Latour, Bruno, 36, 93, 97, 162, 185; parliament of things and, 127
Le Corbusier, 33–34, 89
Lefebvre, Henri, 159
Levenda, Anthony, 263n92
Levine, Philip, 50, 56, 64, 67, 75, 76, 78; on sea-level rise, 54–55
Levy, David, 32
Lexington Insurance, 67
liberalism, 25, 34, 127
Liberty City, 94, 95, 96
Liberty Square, 94
Lilypad (Callebaut), 157
limestone, 41, 147; sea-level rise and, 22
Line, The, 157
Little Haiti, 82, 93, 96
Little River Cooperative, 105, 141
living laboratories, 15, 50, 56, 77
living with water, 55, 88, 119, 186–92
logistics chains, 24

Long, Joshua, 253n92
Long-Term Ecological Research project, 53
Lopez, Jennifer, 74, 146
Lorimer, Jamie, 73
Lovelock, James, 135
Lower Creek Nation, 187
Luke, Timothy, 127

Machu Picchu, 191
Macron, Emmanuel, 181
Madden, David, 57, 154
Magnusson, Warren, 90
makerspaces, 86
Man and the Biosphere program (UNESCO), 120
management: command-and-control, 16, 113; crisis, 126; disaster, 70, 97; economic, 17; ecosystem, 17; environmental, 16; hydraulic, 118; land, 144; redefining/recalibrating, 99, 170; risk, 129, 130; stormwater, 50, 55, 75; strategies, 122, 175; trap, 118; urban, 36; water, 26, 117, 128; waste, 143, 148; wetlands, 125
mangroves, 10, 43, 75, 120, 128, 132, 138, 145, 146, 151, 155, 165
manufacturing, green, 163
Mariel boatlift, 170
marine transgression, 109–10
Mars, colonization of, 157
Martin, Jon, 12
Marvin, Simon, 158
Marx, Karl, 170
"Mask off" (Biscayne Boulevard à la Future), 1
Maxon Cinema 4D, 199

INDEX

McDonald, Michael, 93
McDonald, Robert, 120
McDuffie, Arthur, 95
McDuffie riots, 95, 170
McHarg, Ian, 121
Mckillop, James, 270n1
McKittrick, Katherine, 151
McNeill, John, 3
Medellín, 135, 161
Meeder, John, 110
Merrifield, Andy, 97, 151, 172
Mestas, Juan, 101
Miami: as Anthropocene city, 1–2, 12; as capital of cool, 77, 171, 196; as climate change ground zero, 33, 35, 44, 49, 56, 57–58, 61, 64, 201; as global city, 57; as living laboratory, 15; as Magic City, 58; as Murder City, U.S.A., 170; as Paradise Lost, 170, 171; as urban agglomeration, 11; as Wall Street South, 58, 61, 77
Miami (Didion), 43
Miami Beach, 35, 63, 66, 74, 81, 96, 118; depicting, 199; development of, 43; elevated roads in, 56, 65, 72, 149; infrastructure of, 53, 54, 56, 67, 73, 77, 78, 138; islandization and, 146, 161; marketing campaign by, 55; projection for, 199; promotional stunts on, 45 (fig.); resilience in, 52, 61, 62, 70, 71, 76, 78, 138; sea-level rise and, 46, 50–51, 55; topography of, 146; tourism and, 61

Miami Beach Convention Center, 59
Miami Beach Fire Department, 10
Miami Beach Rising Above, 35, 51, 52, 54, 55, 67, 71, 113, 136; launching of, 64
Miami Bull statue, 59 (fig.)
Miami Climate Alliance, 98
Miami-Dade College, 93
Miami-Dade County, 46, 82, 84, 96, 106, 119, 142; living-with-water toolkit of, 50; resilience in, 15, 62; sea-level rise and, 49, 62; tax collection for, 60
Miami-Dade County Board of County Commissioners, 47
Miami-Dade County Department of Cultural Affairs, 270n1
Miami-Dade Urban Development Boundary, 129
Miami-Dade Urban Long Term Research Area (ULTRA), 240n28
Miami Downtown Development Authority, visualization for, 119 (fig.)
Miami Floating House, 141, 142 (fig.)
Miami Forever, 76–80, 137
Miami Gardens, supply chain and, 106
Miami Herald, 46, 53, 56, 74
Miami Vice, 1
MiamiCoin, 58, 59
Miccosukees, 49, 113, 186; autonomy for, 187; Everglades

and, 116; tree islands and, 187–88
mining sector, 7
mitigation, 31, 52, 66, 70, 144, 151
money laundering, 11
Monk, Daniel Bertrand, 162, 192
Monnin, Alexandre, 40, 160, 193, 194, 195
Moody's, 63, 64, 65, 66
MoonPay, 75
Moore, Jason, 4
Moreno, Gean, 84, 196, 201
Morton, Timothy, 162, 180
Moses, Robert, 89, 91, 121, 129, 241n40
Most Glamorous Cities, 58
Mowry, Bruce, 75, 76, 145–46
Murley, James, 115
Mushroom at the End of the World, The (Tsing), 127
Musk, Elon, 157

Nancy, Jean-Luc, 154
National Academy of Sciences, 132
National Audubon Society, 125
National Oceanic and Atmospheric Administration (NOAA), 46
National Park Service, 125
"Natura Urbans, Natura Urbanata" (Adams), 109
nature, 128; city and, 37, 129–33; Holocenic, 131; infrastructural, 115–24, 127, 129, 130; neoliberal, 120; reengineering, 129; urbanizing, 32, 119
Nature Conservancy, 120
negative commons, 193, 194

Negri, Antonio, 194
Nelson, Sara, 120
neoliberalism, 7, 31, 98, 101, 102, 105, 107, 108, 127, 141, 153, 166
NEOM, 157
networks: citizen, 97; communication, 27; decentralized, 155; delinking from, 163; ecological, 144; energy, 14, 38, 121; fiber-optic, 163; food-production, 106; fragility of, 52; infrastructural, 140, 153, 154, 156, 158, 171, 173; integrated, 173; political-economic, 150; preparedness, 97, 98; production, 38; socio-ecological, 4; transportation, 167; underground, 188; urban, 137; wireless mesh, 106
New Florida Majority, 93
New Orleans, 17; Anthropocene and, 9; post-Katrina, 118
New World Center, 199, 201
New York Times, 42, 46
New Yorker, 44
"Next World, The" (Lovelock), 135
Neyrat, Frederic, 92
Nietzsche, Friedrich, 165
NOAA. See National Oceanic and Atmospheric Administration
NOAA Sea Grant, 142
nonhumans, 131, 172; autonomy of, 132; depictions of, 127; thing-powers of, 127
North American Vertical Datum (NAVD 88), 51

North Beach, 51, 178
North Beach Village, 142, 143, 144 (fig.)
Norton, Glenna, 74
Novo, 60
Nuttle, William, 126

O'Callaghan, Cian, 40
Occupy Sandy, 100
Ogden, Laura, 37, 113, 188–89, 253n77; Everglades and, 128
OKCoin, 59
Olstrom, Elinor, 194
O'Neal, Shaquille, 146
Opa-Locka, 93
Oprah, 161
Order of Importance (Erlich), 48 (fig.)
Overtown, 82, 93, 94, 95, 240n25
Ovink, Henk, 55, 138
oysters, 53, 122, 127

Pacific Rim (film), 199
Palm Island, 51, 146
Paprocki, Kasia, 8, 32, 71, 150–51
Papua New Guinea, 193
Paris, infrastructure of, 131
Parkinson, Randall, 110
Paulson, Morton C., 78
PayPal, 60
Pentagon's New Map, The (Barnett), 153
Pérez, Jorge, 63
Pérez Art Museum Miami (PAMM), 63
permafrost, melting, 172
Peters, Kimberley, 165
Phoenix, climate change and, 190

planning: approaches, 18, 23; central, 191; climate change and, 3; decentral, 165, 191; engineering-based, 145; exercises for, 87; modern, 91, 112; resilience, 84, 91; transformative, 190; workshops, 139. See also urban planning
Plato, 30
"Point and Pivot" (Zuluaga), 1
politics, 31, 184; corruption in, 96
pollution, 154, 160, 187
poverty, 82, 95, 96, 98, 135
Praxis Society, 157
preservation, 115, 175, 195
Price, Rene, 129, 132
problematizations, 16, 19–20, 21, 23, 28, 30, 33, 49, 112, 117, 177; context-specific, 34; imaginaries and, 39; resilience, 16, 99; urban Anthropocene, 115
proletarianization, 163
Prologue: Politics as Palliative Care of the Species (film), 140
property values, 56, 64, 74
"Protocols for the Phase Transition" (A.S.T.), 141, 183
Public Works Administration, 94
Pugh, Jonathan, 155
pumps, 35, 50, 51, 56, 71, 72, 75, 113, 117, 118, 136, 145

Rabinow, Paul, 21
Rabois, Keith, 60
race, realities of, 196
rainfall, flood-inducing, 12
rainwater, 143; harvesting, 106, 179, 190

RAND Corporation, 85
real estate, 35, 55, 98, 149, 199, 200; boom, 78; crash, 202; development, 43, 56; investment in, 57; market, 59, 61, 75, 77, 92; sea-level rise and, 14; speculation, 57, 78; values, 62, 65
RealtyHop, 60, 84
Rebuild by Design, 55, 91
Redlands, 88, 93, 96, 106
reefs, wave-attenuating, 127
Regional Refuge, 155
rehydration, 125, 188
Reid, Julian, 28, 36, 92, 102, 103, 104
relationships: human-environment, 165, 185, 253n77; human–nature, 30; social-ecological, 166
relocalization, 140, 158
relocation, 116, 161, 187
reproblematization, 25, 57–58, 70
Republic, The (Plato), 30
Republican National Convention, 95
resilience, 2, 7–8, 10, 20, 34, 39, 52, 98, 156, 171, 183, 192; allegiance to, 198; approaches to, 17, 122; assumptions about, 23; building, 9, 18, 64, 65, 84, 120, 145, 189; characterizations of, 126–27; climate, 63, 65, 66, 71, 100, 151; as crisis response, 61–67; economies of, 56–60; extreme adaptation and, 137–39; goals of, 91–92; imagination and, 177; impact of, 65, 73; interventions, 76; logics, 175; practice, 107–8; projects, 50, 56, 104, 213n33; protective, 55; realities of, 76; reappropriating, 191; sea-level rise and, 66; survival and, 183; sustainability and, 16; urban geographies and, 145; urban governance and, 17. See also urban resilience
Resilience Age, The (Rockefeller Foundation), 135
Resilience Alliance, 17
Resilient 305, 15, 84
resilient city, 37, 38, 102–3, 113, 137, 173; paradigm of, 136
Resources for the Future, 14
restoration, 115, 120, 125; restitution and, 126
rice farming, 150
Rice, Jennifer, 263n992
Rickards, Lauren, 40
Rio Tinto Zinc, 193
rising above, 49–52, 75
risk, 129, 130; knowing, 43–44, 46–49
risk rent, 98
Risky Business Project, 61
robotics, 57, 106
Rockaway Beach Surf Club, 100
Rockefeller Foundation, 50, 86, 135, 145
Rockström, Johan, 72
Rolling Stone, 49, 54
Roman, Michael, 150
Rose, Gillian, 190
Rose, Nikolas, 101
Roseate spoonbills, 114, 132
Royal Library of Alexander, 155
RTZ corporation, 193
Russia-Ukraine War, 156, 158

Said, Edward, 32
saltwater, 128, 133, 151; corrosion from, 172; intrusion, 22, 109–10, 114, 115, 116, 130, 131, 132, 136, 150, 189
saltwater railroad, 188
Salty Urbanism, 142, 155, 158
Samoylova, Anastasia, 13
Sánchez-Calderón, George, 104; house fire by, 81–82, 83 (fig.)
Scarface (film), 44
scenario visioning, 85, 100, 190. See also futures scenarios
Schwartz, Peter, 81
Scott, Allen, 167
Scott, James C., 89, 90
Scott, Rick, 15, 139
sea-level rise (SLR), 15, 16, 37, 41, 51, 54–55, 71, 72, 78, 98, 114, 116, 128, 132, 136, 141, 144; adaptations for, 2, 10; assumptions about, 66; combating, 127; economic threats from, 62–63; as fake news, 60; fighting, 171; flooding and, 12, 44, 62; impact of, 2, 14, 172; increase in, 13–14, 42; limestone and, 22; managing, 127; maps, 42, 47, 201; permanent, 61–62; porous substrate and, 46; projections for, 48, 50, 76, 138, 143; real estate and, 14; resilience and, 66; risks of, 46, 49, 115; tourism and, 61–62; vulnerability to, 109, 146
Sea Level Rise Strategy (SLR Strategy), 49, 50, 51, 214n42
Sea Level Solutions Center, 64
Seasteading Institute, 157
seawalls, 8, 46, 56, 118

security, 97; climate change and, 3; climatization of, 70; ecological, 177; threats, 85; urban, 91, 137, 158
segregation, 44, 82, 240n25; impact of, 22; legacies of, 95; racial, 11
Seminoles, 49, 113, 114 (fig.); Everglades and, 116
September 11th, 26, 153
Serious Games, 93, 95–96, 98, 103, 104, 105
sewage systems, 110
Shake Shacks, 8
Shell, 85
Shelton, Taylor, 30
Shorecrest, 93
shrimp aquaculture, 8, 150
Silicon Valley, 35–36, 58, 157
Simone, AbdouMaliq, 36, 99, 101
Sloterdijk, Peter, 165
SLR. See sea-level rise
SLR Strategy. See Sea Level Rise Strategy
slums, 26, 150
social conditions, 73, 145
social conflict, 2, 30, 78, 119
social infrastructure, 23, 36, 40, 98–103, 136, 137, 145, 191; embracing, 113; reappropriating, 103–8
social life, 52, 101
social media, 44, 58
social relations, 101, 153, 253n77
social systems, 70, 71, 73
socio-ecological change, 85, 107, 185, 196
socio-ecological systems, 8, 17, 27, 53, 56, 79, 86, 89, 90, 96, 113, 121, 175, 180

socioeconomic conditions, 111, 123
socionatures, Anthropocenic, 37, 129, 131
Soft Defense (The Green Jacket), 143
SoftBank, 60
Soja, Edward, 12, 167
Solar Atlas, 72
Soleri, Paolo, 159, 160
Soto-Wright, Ivan, 75
Souriau, Étienne, 194–95
South Florida, 43, 46, 96, 118, 146; agriculture in, 109; island imaginaries of, 140–45, 154; migration from, 139, 147; sea-level rise and, 147
South Florida Climate Resilience Tech Hub, 15
South Florida Serious Games Disaster Preparedness Exercise, 93
South Florida Water Management District, 111–12, 125
Southeast Florida Regional Climate Change Compact, 48, 61
space, 8, 25; controlling/organizing, 263n104; problem, 21; urbanized, 172. See also urban spaces
SpaceX, 157
Spicer, André, 32
Star Island, 146
state sight, 89, 90, 93, 162; Anthropocene, 36, 91, 101, 102
Steffen, Will, 3
Steinberg, Phil, 165
Stengers, Isabelle, 185

Stiegler, Bernard, 3, 186
storm surge, 46
Strategic Retreat, 143
strategies, 88, 98, 184, 197; adaptive, 9, 161, 172, 191; amphibious, 144; depresencing, 195; governance, 84; infrastructural, 130, 156; management, 122, 175; survival, 184; synthetic, 185
Suarez, Francis X., 58, 63, 78; tweet by, 59 (fig.)
Sunset Harbour Publix, 51
supply chains, 26, 106, 156
surveillance, 26, 57, 163, 191
survival, 102; infrastructure and, 164; resilience and, 183; skills, 106, 179, 189
sustainability, 17, 79, 89, 97, 113–14, 122, 158, 160; resilience and, 16; solutions, 107
Swamp Life: People, Gators, and Mangroves Entangled in the Everglades (Ogden), 188
Swamp Peddlers, The (Vuic), 78
Swyngedouw, Erik, 4, 31, 104

Taljanky, 197
Tamiami Trail/US-41 highway, 125–26
Taylor, Zac, 39, 98
technodiversity, 186
technology, 29, 30, 36, 58, 112, 117, 130, 186, 187, 203; advanced, 165; assemblages of, 25; computation, 32; governmental, 123; high/low, 182; ordering, 128; resilience, 122; solar, 155; tracking, 57
telos, 154, 172

Telosa City, 157
Teotihuacan, 197
terraforming, 140, 146, 165
Thiel, Peter, 60, 157
thing power, 128, 129
thinking: Anthropocenic, 179, 180, 181–82, 184; complex systems, 121; critical, 84, 99, 190–91; development, 16; ecology, 16; future-oriented, 192; resilience, 99; strategic, 192. See also urban thinking
Thrift, Nigel, 90
Thunberg, Greta, 139
Time, 58, 170, 171
tourism, 55, 60, 64, 72, 94, 119, 148, 155, 169; circulation of, 136; dependency on, 11; sea-level rise and, 61–62
traffic jams, sand-carved, 48 (fig.), 49
transformation, 13, 23, 27, 29, 72, 79, 80, 88, 90, 92, 131, 139, 141, 167; economic, 31; emancipatory, 102–3; environmental, 6, 31, 68; infrastructural, 155; political, 6; possibilities for, 37; social, 6, 31; spatial/institutional/ecological, 6; urban, 5, 10, 164
transportation, 121, 156, 167, 186
tree islands, 187–88
Tropical Malaise: Prologue (film), 199, 201; stills from, 200 (fig.)
Truscello, Michael, 37
Tsing, Anna, 127, 180, 191
Turkey Creek Nuclear Plant, 93
2 Fast 2 Furious (film), 44

ULI. See Urban Land Institute

"Unified Sea Level Rise Projection for Southeast Florida," 48
United Nations Education, Scientific, and Cultural Organization (UNESCO), 120
Unplanning Miami, 143, 145, 148, 149, 155, 164, 165, 194
Upper North Bay Road, elevations on, 73–74
urban age, 1–5, 150
urban condition, 5, 39, 153
Urban Condition, The (Gleeson), 135
urban depresencing, 192–95
urban design, 22, 85, 176, 182
urban development, 37, 43, 62, 82, 94, 103, 143, 148, 171; wetlands and, 115–16
urban dismantling, 145–48, 160
urban-ecological systems, 154, 159–60
urban form, 22–23; experimentation in, 195–98
urban futures, 28, 30, 92, 104, 105, 136; imagining, 2, 15, 84–86; improved, 72
urban geography, 42, 145, 160, 169
urban governance, 25, 52, 57, 69, 97, 101, 118, 176; approaches to, 16, 119; complexity science and, 90; Everglades and, 112; problematization of, 16, 27, 112, 117; recalibrating, 99–100; resilience and, 17; transformations of, 7, 68
urban infrastructure, 19, 22, 24–35, 36–37, 121, 124; critical, 116, 122, 154; problematization of, 123; redefining,

27; remodulations of, 23, 24; securing, 122; threats to, 116; transformations of, 23
Urban Land Institute (ULI), 14, 55, 61, 62, 118–19
urban life, 6, 40, 77, 92, 104, 110, 137, 154, 177, 182, 196; climate change and, 112; infrastructure and, 26
urban planning, 22, 27, 30, 31, 34, 89, 127, 136, 176, 182; approaches to, 21; critiques of, 23; ecological, 120; engineering-based, 121; modern, 85, 117
urban regime, 153, 169, 193
urban resilience, 5, 8, 9, 10, 16–18, 19, 21, 23, 32, 36, 38, 39, 50, 53, 73, 120–21, 122, 126, 135–36; Anthropocene and, 28; critical interrogation of, 7; discourses of, 190; promise of, 173; repetition of, 164; understanding of, 33; urbanization, 166–67; wetlands and, 116
Urban Resilience to Extreme Events Simultaneously Research Network (UREx-SRN), 87
urban risks, 22, 113, 123
urban spaces, 6, 9, 30, 31, 92, 103, 143, 150; mutations of, 4–5; recalibration of, 2; resilience of, 192; understanding of, 32
urban studies, 19–23
Urban Studies Foundation, 24
urban systems, 4, 9, 69, 102, 103, 121, 123, 223; ecocybernetic, 18; globalized, 171–72; human-natural, 122; impact of, 23; interconnected, 6; managing, 16; preserving, 175; social-ecological, 85; socioeconomic, 36, 191
urban theory, 6, 40, 42, 173–78
urban thinking, 33, 149, 173, 177, 196; critical, 10, 195; Holocene, 40; inherited, 153; pushing beyond, 6
urbanization, 2, 7, 8, 17–18, 31, 32, 38, 102, 148, 151, 152, 153, 163; assumptions about, 23; back loop, 35, 67–73, 76; capitalist, 5–6, 26, 57; development of, 4, 13, 157, 160; global, 137, 154, 167, 172, 173; littoral, 188; naturalizing, 70; networked, 172; planetary, 9, 10, 39, 106, 149, 151, 152, 154, 158, 159, 166, 170, 171, 172, 174, 185; processes of, 18, 23, 33; replacing, 173; resilient, 90, 91, 137, 166–67; vocabularies of, 6
Urbicene, 4
urbicide, 38–39, 149–52, 156, 158, 164, 165, 171, 173
U.S. Army Corps of Engineers, 75, 112, 113, 119, 125

Valle-Levinson, Arnoldo, 12
Van Munster, Rens, 70
Vaughan-Williams, Nick, 76
vegetation, saltwater-resistant, 143, 155
Virginia Key, sea-level rise and, 46
visualization, 30, 31, 33, 119, 201

Vuic, Jason, 78

Wachsmuth, David, 137–38
Walsh, Virginia, 189
Wanless, Harold, 49, 110, 138; diagram by, 47 (fig.); maps by, 47, 49
water: budget, 80; defending against, 145–46; drinking, 107, 111, 117, 121, 126, 131, 190; flows, 111, 113; purification, 140; quality, 113, 114; shortages of, 189, 190
Water Knife, The (Bacigalupi), 190
water systems, 202; contamination of, 14, 49; subtending/surrounding Miami, 110 (fig.)
Wengrow, David, 97

West 8's Soundscape Park, 199
West Coast, wildfires on, 82
wetlands, 43, 111, 125; urban development and, 115–16
Wheaton, Elizabeth, 55
wilderness skills, 106
wildfires, 2, 82
wildlife corridors, 140
WLRN, 46
Work Group, 48

Years of Living Dangerously (Black), 55
Yusoff, Kathryn, 32, 79

Zalasiewicz, Jan, 4
zone of transit, 203
Zoom, 58
Zuluaga, Natalia, 1, 200, 201

Stephanie Wakefield is assistant professor of urban planning and environmental design, Department of Urban and Regional Planning, Florida Atlantic University.

www.ingramcontent.com/pod-product-compliance
Lightning Source LLC
Chambersburg PA
CBHW071617130325
23411CB00004B/24